Department of the Environment

Asbestos materials in buildings

LONDON: HMSO

© *Crown copyright 1986*
First published 1983
Second edition 1986
Third edition 1991

ISBN 0 11 752370 4

Contents

Section *Paragraph* *Page*

1 Introduction   5

    The nature of asbestos   1.5   5
    The Advisory Committee on Asbestos   1.7   6
    Legislation on asbestos in buildings   1.9   6
    Legislation to protect the public   1.11   7
    The Building Regulations and British Standards 1.16   8
    The European Market   1.18   8

2 Health effects   9

    Asbestos related disease   2.2   9
    Exposure, latency and fibre size   2.3   9
    Asbestos in buildings   2.5   9
    The public health risk   2.9   10
    Asbestos in drinking water   2.15   11

3 Asbestos materials used in buildings   13

    Sprayed coatings and lagging   3.2   13
    Insulating boards   3.5   14
    Ropes, yarns and cloth   3.7   14
    Millboard, paper and paper products   3.8   14
    Asbestos-cement products   3.9   14
    Bitumen felts and coated metal   3.17   16
    Flooring materials   3.19   16
    Textured coatings and paints   3.20   16
    Mastics, sealants, putties and adhesives   3.21   16
    Reinforced plastics   3.22   16

4 Identification and assessment of existing asbestos installations 17

    Identification   4.2   17
    Assessment   4.5   17
        The potential for fibre release 4.6   17
        The type of material   4.7   18
        The integrity of the material   4.10   18
        The position of the material   4.11   18
    Measurement of airborne asbestos fibre concentrations 4.12   18
    Quality control and selection of laboratories   4.16   19

Section	Paragraph	Page
5 Remedial measures, surveys and priorities		20
Remedial measures and management	5.3	20
Management	5.4	20
Sealing and repair	5.5	21
Enclosure	5.6	21
Removal	5.7	21
Surveys	5.9	21
Priorities	5.13	22
6 Demolition, disposal and fires		23
Demolition, including modifications and refurbishment	6.1	23
Disposal of asbestos waste	6.2	23
Asbestos in fires	6.10	24
7 Appliances and equipment containing asbestos		26
Domestic appliances	7.3	26
Household goods	7.5	26
Asbestos fire blankets	7.6	26
Asbestos in heating systems	7.7	27
Electric 'warm air' and storage heaters	7.12	28
Do-it-Yourself – DIY work	7.14	29
Labelling of asbestos products	7.18	29
8 Substitutes for asbestos		30
Sprayed fire protection	8.4	30
Thermal insulation and lagging	8.5	30
Insulating boards	8.7	30
Asbestos-cement	8.8	31
Other products	8.11	31
Health aspects of mineral fibre substitutes	8.13	32

Annexes

	Page
1. Asbestos materials used in buildings	34
2. Sampling asbestos materials	39
3. Asbestos assessment charts	40
Chart 1: Asbestos materials	40
Chart 2: Sprayed asbestos and lagging	42
Chart 3: Asbestos insulating board, insulating blocks and composite products	44
Chart 4: Asbestos-cement products	46
4. Notes on procedures for other asbestos materials	48
5. The asbestos label	49
6. Model asbestos survey report form	50
7. The asbestos waste label	51

References	53

Tables and Photographs

Table 1 Results of airborne asbestos monitoring	page 12
Photographs	between pages 28 and 29

1 Introduction

1.1 This third edition of Asbestos Materials in Buildings has been prepared to take into account technical developments in the field of asbestos and substantial new legislation which has been introduced since 1986 when the booklet was last revised. The first edition, published in 1983, presented in a single publication descriptions of the main past and present uses of asbestos in buildings and gave general advice on the precautions which should be taken when dealing with asbestos materials. The second edition which took into account the findings and recommendations of the Joint Central and Local Government Working Party on Asbestos, updated the information on controls on the use of asbestos materials, presented the results of further studies of the effects on health of asbestos, and gave more detailed guidance on assessment of the risk from asbestos in buildings. Asbestos has been used in building materials for many years and there is a large tonnage of asbestos materials in existing buildings. The use of asbestos is new building materials has been sharply reduced in recent years but a significant tonnage is still used, mainly in the manufacture of asbestos-cement. A considerable volume of information and advice on asbestos is available in a variety of books, reports, advice notes and other publications. Sources of more detailed information and advice are listed at the end of the booklet.

1.2 The advice is aimed at all those with responsibilities for buildings including local authority staff responsible for public buildings and housing, environmental health officers and others called upon to advise the general public, safety representatives, and maintenance engineers responsible for public and private buildings. Local authorities have general responsibilities under public health legislation where premises are prejudical to health and are also responsible for building control and asbestos waste disposal. As employers they have responsibilities under health and safety legislation, and are responsible for enforcing the Health and Safety at Work etc. Act 1974 and the Control of Asbestos at Work Regulations 1987 (paragraph 1.10) in many premises where work with asbestos may be carried out. As housing authorities they have responsibilities to ensure the safety of their tenants and also have the responsibility of advising private householders and taking action if necessary. As education authorities, they are responsible for providing school buildings of such design and contruction that the health and safety of the occupants is reasonably assured. They provide and maintain a wide range of buildings for their various services.

1.3 When advising the general public in particular it must be recognised that a considerable degree of anxiety over asbestos-containing materials in buildings is still prevalent. It is essential therefore to emphasise that there is no risk to human health from the simple presence of such materials if left alone and undamaged. It is only when respirable fibres are released that risk becomes apparent, and then only to a degree as described in Section 2. When carrying out work on any fibrous materials, there is a need therefore for awareness, precaution to control dust levels, the maintaining of hygiene standards and the following of good working practices as set out in this booklet. This philosophy must be reflected not only when such work is carried out by local authority workers and others, but when advising the general public to whom copies of the leaflet 'Asbestos in Housing'[39] should be distributed freely. The public should also be made aware of the general safety precautions recommended by the Health and Safety Executive (HSE) and guidance given in paragraphs 3.12–3.16.

1.4 This booklet does not cover specialised technical uses of asbestos in factories, power stations or production plant. Nor is it intended to give detailed advice on the protection of construction, demolition or maintenance workers exposed to asbestos. It is important that the correct safety precautions are taken when working with asbestos materials and all such work is subject to the Health and Safety at Work etc. Act 1974 and associated legislative requirements in particular, the Control of Asbestos at Work Regulations 1987. Detailed information and advice on work with asbestos can be obtained from HSE local offices in the form of Guidance Notes, Approved Codes of Practice and Leaflets etc. The Abestos Removal Contractors Association also publish asbestos related publications.

The nature of asbestos

1.5 Asbestos is a fibrous mineral which occurs in many parts of the world; the main sites of commercial production are in Canada, the Soviet Union and Southern Africa. The three main types of asbestos produced commercially are:

Crocidolite – blue asbestos
Amosite – brown asbestos
Chrysotile – white asbestos

About 6 million tonnes of all types of asbestos have been imported since the turn of the century. Levels peaked in 1973 at approximately 195,000 tonnes, fell in 1984 to less than 40,000 tonnes, and in 1989 to 25,000 tonnes. Imports of crocidolite and amosite ceased in 1972 and 1980 respectively.

1.6 Asbestos-containing rock is crushed and milled at the mining site to produce raw asbestos of various grades. Asbestos fibre is mechanically strong and highly resistant to heat and chemical attack and, because of its fibrous nature, it can be woven into fabrics and used as reinforcement for cement and plastics. It is the very fine fibres, invisible to the naked eye, which are dangerous when inhaled, and processes which produce very small airborne fibres are, in general, the most hazardous. These fibres pose no threat to health when left intact in undamaged material containing absestos, which should be either left alone and managed (periodically assessed for deterioration) (paragraphs 5.3 and 5.4), sealed or enclosed (paragraphs 5.5 and 5.6) or removed if deteriorating (paragraph 5.7). In order to determine the state of the material local authorities will need to arrange for surveys to be carried out (paragraphs 5.9–5.12) in their buildings and to establish an order of priority for any measures which are to be undertaken (paragraphs 5.13–5.16).

The Advisory Committee on Asbestos

1.7 In 1976 the Health and Safety Commission (HSC) appointed an advisory committee (The Advisory Committee on Asbestos – ACA) to look into risks to the health of workers and the general public arising from exposure to asbestos. The Committee reported in 1979[1] making a number of recommendations for further action to reduce exposure to asbestos. On asbestos in buildings the Committee said:

> 'Firm conclusions about asbestos dust levels in buildings cannot be drawn from the present data. However . . . the number of people at risk is probably small'.

and:

> 'Our evidence of the non-occupational risk is not such as to prompt us to recommend the general removal of asbestos from buildings. Present evidence suggests that dangers from asbestos in buildings are likely to arise only when asbestos fibres are released into the air when products containing asbestos are damaged, either accidentally or during maintenance and repair'.

1.8 The Committee recommended that further monitoring of the levels of airborne abestos should be carried out. Results of this work are now available and are discussed in the following section of this booklet. They support the Committee's conclusions on the risk from asbestos in buildings. The ACA also made a number of other recommendations about asbestos used in building materials. Details of how some of these recommendations have been implemented are set out in paragraph 1.10.

Legislation on asbestos in buildings

1.9 The Health and Safety at Work etc. Act 1974 (Chapter 37) places general duties on employers and self employed persons to ensure as far as is reasonably practicable, the health, safety and welfare at work of all their employees and persons other than their employees who may be affected by any of their undertakings (this may include working with asbestos); they must also ensure that the premises and any plant or substance therein are safe and present no risks to health from substances such as asbestos.

1.10 Regulations have also been introduced under the Health and Safety at Work etc. Act 1974, the Control of Pollution Act 1974, the Water Act 1989 and the Consumer Safety Act 1978 (Chapter 38) to implement several of the recommendations of the ACA and to comply with the requirements of four European Communities' directives on asbestos. [3, 4, 45, 46].

- The Asbestos (Licensing) Regulations 1983 (Statutory Instrument 1983 No. 1649 which came into effect on 1 August 1984) prohibit work with asbestos insulation or coating, by an employer or self-employed person without a licence granted them by the HSE[5]. Short-term repair or maintenance work, or work in premises occupied by the employer or self-employed person are allowed under the Regulations provided advance notice of the work has been given to the enforcing authority.

- The Asbestos (Prohibitions) Regulations 1985 (Statutory Instrument 1985 No. 910 which came into force on 1 January 1986) prohibit the importation, use in manufacture and marketing of crocidolite and amosite and products containing them, asbestos spraying and asbestos insulation.

- The Asbestos (Prohibitions) (Amendment) Regulations 1988 (Statutory Instrument 1988 No. 711 which came into force on 25 April 1988) introduce further prohibitions on the supply of products intended for spraying and the supply and application of paints containing asbestos for use at work.

- The Asbestos Products (Safety) Regulations 1985 (Statutory Instrument 1985 No. 2042 which came into force on 1 January 1986) prohibit the supply of crocidolite and amosite and products containing them and, from 20 March 1986, require the labelling of products containing asbestos.

- The Asbestos Products (Safety) (Amendment) Regulations 1987 (Statutory Instrument 1987 No. 1979 which came into force on 1 January 1988) ban the retail supply of 6 types of asbestos product: toys, products for spraying, products in powder form intended for private use or consumption where their supply is (or would be by retail), items for smoking, catalytic gas heaters, and paints and varnishes.

- The Control of Asbestos at Work Regulations 1987 (Statutory Instrument 1987 No. 2115 which came into force on 1 March 1988) apply 'when any work with asbestos or with any product containing asbestos is carried out by the employer'. They require that an employee's exposure to asbestos be prevented or reduced as far as is reasonably practicable, and set down for the first time in legislation control limits at or above which employees must not be exposed unless they are wearing respiratory protective equipment. Employers are obliged to assess any risk to employees before they carry out work with asbestos so that a correct decision can be made about the measures necessary to control exposure. If the assessment indicates that the cumulative exposure measured over a continuous 12-week period will, or is liable to, exceed the prescribed action level, then other provisions of the Regulations are triggered. These include notifying the enforcing authority of the work concerned, measuring the amount of asbestos in the air in the workplace, posting warning signs in work areas, health surveillance and maintenance of health records. The duties imposed on employers to protect their employees are extended to anyone else who may be affected by the work, including members of the public. Also, there is a specific requirement that the spread of asbestos from any place where work with it is carried out is reduced as far as is reasonably practicable.

- There are 2 Approved Codes of Practice (ACoP) which operate when work is carried out with asbestos. The first provides practical guidance on complying with the Control of Asbestos at Work Regulations 1987[6], whilst the second ACoP provides practical guidance on work with asbestos insulation, asbestos coating and asbestos insulating board[2].

- The Control of Asbestos in the Air Regulations 1990 (Statutory Instrument 1990 No. 556 which came into force on 5 April 1990) implements the emission limit of 0.1 mg/m^3 of asbestos to the air by industrial installations. The Regulations also include further general provisions to prevent significant environmental pollution from activities involving the working of products containing asbestos and the demolition and removal of materials containing asbestos.

- The Control of Pollution (Special Waste) Regulations 1980 (Statutory Instrument 1980 No. 1709 which came into force on 16 March 1981) place additional controls over wastes which are particularly difficult or dangerous to dispose off. They allow for tighter control over the carriage of wastes and require a consignment note procedure to be carried out before the waste is moved.

- The Collection and Disposal of Waste Regulations 1988 (Statutory Instrument 1988 No. 819 which came into force on 6 June and 3 October 1988) implement Sections 12–14 of the Control of Pollution Act 1974; they prescribe the descriptions of waste which are to be treated as household, commercial and industrial waste, and set down the cases where a disposal licence is required for the disposal or handling of controlled waste.

- The Trade Effluents (Prescribed Processes and Substances) Regulations 1989 (Statutory Instrument 1989 No. 1156 which came into force on 1 September 1989) enable specific control of discharges of asbestos to public sewers of asbestos derived from processes for the manufacture of asbestos-cement and asbestos paper and board.

Legislation to protect the public

1.11 In general, statutory responsibility for enforcement of work place safety, including construction and demolition sites, will rest with the HSE. The measures required to protect people working on the site from exposure to asbestos will usually also prevent exposure of the public. There is a general responsibility under Section 3 of the Health and Safety at Work etc. Act 1974 and a specific requirement under Regulation 3 of the Control of Asbestos at Work Regulations 1987 to protect the health and safety of the general public who may be affected by work activities. However there may be occasions when the additional legislative powers, referred to in paragraphs 1.12–1.14, may be valuable in tackling a particular problem involving asbestos.

1.12 Certain provisions of the Environment Protection Act 1990 (previously found in the Public Health Act 1936) are relevant to the use of asbestos materials in the construction and demolition of buildings. The Environment Protection Act (Sections 79–83) which replaced Sections 92 et seq. of the Public Health Act 1936) defines a number of statutory nuisances which may be dealt with summarily, including 'any dust or effluvia, caused by any trade, business, manufacture or process and injurious, or likely to cause injury, to the public health or a nuisance'. The Act gives local authorities the power to serve abatement notices where premises are in such a state as to be prejudicial to health, or a nuisance.

1.13   Section 82 of the Building Act 1984 (previously Section 29 of the Public Health Act 1961), gives local authorities powers to impose conditions on the demolition of buildings. The local authority must be informed of the intention to demolish certain buildings and the authority can serve a notice requiring the person carrying out the demolition to meet specified conditions. These include removal of material or rubbish from the site and measures for the protection of the public and preservation of public amenity.

1.14   The Defective Premises Act 1972 (Chapter 35) (England and Wales only) puts an obligation on a person taking on work in connection with the provision of a dwelling to see that the work is done in a workmanlike manner, with proper materials and that the dwelling is fit for habitation when completed, ie. not exposed to asbestos levels above the clearance indicator levels set by HSE[18]. In the case of a property let under a tenancy which puts an obligation on the landlord for the maintenance and repair of the premises, the Act places a duty on the landlord similarly to take reasonable care to see that the tenant and other people are safe from personal injury or disease caused by a defect in the state of the premises.

1.15   Abandoned industrial premises, former waste disposal sites (including those already reclaimed for agricultural or amenity use etc.) and other derelict land may be contaminated by asbestos. In some cases the asbestos may be buried, while at some sites it may be present above ground, eg. on sites where it was used for heat insulation, for fire control, or in the construction of walls and roofs of buildings. Asbestos may be in the form of manufactured arcticles such as asbestos-cement boards, sheets and pipes, or present as loose unconsolidated deposits which can easily give rise to airborne fibres. General advice and guidance on the identification, assessment and treatment of sites contaminated by asbestos, has been published by the Interdepartmental Committee on the Redevelopment of Contaminated Land[44].

The Building Regulations and British Standards

1.16   The relevant national building regulations impose requirements about how buildings are designed and constructed with the objective of protecting public health and safety and the environment. There is no requirement in the regulations to use any materials containing asbestos, but such materials may be used if they fulfil the specified performance standards. For example asbestos-cement flue pipes meet requirements for non-combustibility, and asbestos-cement tiles and sheet meet the requirements for flame spread in fire-resisting construction. The current regulations contain no references to asbestos, but the 'Approved Documents' which give practical guidance about some of the ways of meeting the requirements of the regulations (such as those on Fire and Hygiene) refer to asbestos-cement products for use as slates, corrugated sheet and cavity roof decking, and as pipes for drainage or sewerage. Earlier regulations listed asbestos insulating board as a suitable material for fire-resistant walls and ceilings, and sprayed asbestos as fire-resisting protection for steel framework and other constructions. The use of asbestos insulation or sprayed asbestos is now prohibited by the Asbestos (Prohibitions) Regulations 1985 and 1988 (paragraph 1.10). Although exempted from national building regulation control, maintained educational buildings are covered by Constructional Standards which are closely related to the standards of the relevant building regulations.

1.17   In general, British Standards Institution Technical Committees aim to avoid reference to the use of hazardous asbestos fibres in products and seek to refer to the use of alternative materials. Under the influence of market forces and new and impending legislation, manufacturers of asbestos-based building products are gradually introducing non-asbestos products, but asbestos products such as slates, are likely to be manufactured for a number of years and the relevant British Standard specifications for those products therefore need to be retained for that period. The BSI Fibre Reinforced Cement Products Standards Committee has developed a strategy for amending existing British Standards to incorporate current safety requirements and, at the international level in the International Organisation for Standardisation (ISO), of supporting the development of the second generation of products based on compositions which exclude asbestos.

The European Market

1.18   The Construction Products Directive (89/106/EEC) is designed to promote free trade within the EC. The principal effect of the Directive will be that products bearing the CE mark are deemed to be 'fit' for sale. In practice 'fitness' will commonly be demonstrated by the product conforming to a harmonised European standard. The fitness of a particular product for use in works unregulated by national provisions will still be dependent on the product satisfying the requirements of relevant national or local legislation. However, such legislation must itself be compatible with the requirements of the Directive. If, for example, a harmonised European Standard provided for a class of products that did not contain asbestos, it would be perfectly possible for a Member State to allow only products that did not contain asbestos to be used in works regulated by national provisions. In producing harmonised European Standards CEN would, in any event, have to take due account of the requirements of other Community Directives including in particular, the requirements of Marketing and Use Directives. The effect of regulations made under these Directives is such that residual trade within the Community in products incorporating asbestos is likely to be extremely limited.

2 Health effects

2.1 A review of the available scientific evidence on the effects of abestos on human health was carried out for the Advisory Committee on Asbestos (ACA) by Acheson and Gardner and published in the report of the Committee in 1979.[1] An update of sections of the report was made by Acheson and Gardner at the request of the HSC and published in 1983.[7] A further review of the adverse effects of asbestos on health was prepared by Doll and Peto for the HSC and published in 1985.[8] Major studies of the health effects of asbestos were also carried out in the USA[9,10] and Canada[11] during this period. The report of an IPCS Working Group held in Rome in December 1988, contained the collective views of an international group of experts, which supported the broad agreement which there is between these studies, on the effects of exposure to low concentrations of asbestos in the environment. The general conclusions of these studies are summarised in this section.

Asbestos related diseases

2.2 The principal diseases known to be caused by exposure to asbestos are asbestosis, lung cancer and malignant mesothelioma.

- Asbestosis. Fibrosis or scarring of the lung in which the tissue becomes less elastic making breathing progressively more difficult. It is irreversible and may progress even after cessation of exposure to asbestos. Asbestosis is an industrial disease arising from high levels of exposure to airborne dust and there is no risk of contracting this disease from normal levels of environmental exposure to asbestos.
- Lung cancer. An increased incidence of lung cancer has been found amongst people who work with asbestos. The increase in risk depends on the degree of exposure and is very much greater for smokers than for non-smokers. All three types of commonly-used asbestos fibre can cause lung cancer, but crocidolite and amosite are thought to be more dangerous than chrysotile.
- Mesothelioma. A cancer of the inner lining of the chest or of the abdominal wall. The incidence in the general population is very low: the overwhelming majority of cases are attributable to occupational or para-occupational (those living in the same house as an asbestos worker) exposure to asbestos.

Exposure, latency and fibre size

2.3 The risk of contracting an asbestos related disease depends on a number of factors including the cumulative dose to which an individual has been exposed, the time since first exposure and the type of size of the asbestos fibres. It is generally assumed that the risk of cancer is proportional to total exposure, but according to Doll and Peto,[8] the risk of mesothelioma is also strongly related to the time since first exposure. There is commonly a lag or latency period of 10-20 years between first exposure and onset of symptoms for asbestos related diseases and, in the case of cancer, the period of latency may be up to 40 years or more.

2.4 Fibre size and shape are thought to be important variables in determining the risk from asbestos. Longer fibres with a length of greater than 200 um (1 um = one millionth of a metre) are generally cleared from the nasal passages, but shorter fibres with a diameter of less than about 2 um may penetrate deep into the lungs. Doll and Peto[8] summarise the evidence thus:

> 'Laboratory evidence suggest that the hazard is greatest with fibres between 5 and 10 um in length and less than 1.5 or 2 um diameter. There are, however, no sharp boundaries between hazardous and non-hazardous configurations. Short fibres less than 1 or 2 um in length may not be hazardous at all; but there is no evidence of any minimum diameter to hazardous fibres which may be carcinogenic even when the diameter is so small that they cannot be seen by the optical microscope.'

Asbestos in buildings

2.5 The ACA recommended that a programme of work should be started to evaluate asbestos exposure in the non-occupational environment, and a programme, jointly managed by the Department of the Environment (DOE) and HSE, was set up in 1978. Measurements have been made in a wide range of buildings in which asbestos materials were present.

As far as possible samples were taken during normal use of the buildings using large volume air samplers, and the filters were analysed by optical and transmission electron microscopy.

2.6 The samples obtained in the initial stage of the monitoring programme were analysed in terms of the mass of airborne asbestos and the results have been published[12]. Measurements were made in ten buildings of various types, including schools and houses, and in no case was the concentration of airborne abestos above the limit of quantification of the method used (10 nanograms per cubic metre (ng/m^3) for scanning electron microscopy and 1 ng/m^3 for transmission electron microscopy. 1 ng is one thousand millionth of a gram).

2.7 Subsequent measurements were made in terms of fibres per unit volume and the results are summarised in Table 1. They are reported in terms of fibres greater than 5 um in length, less than 3 um in diameter and with an aspect ratio (length to diameter ratio) of greater than 3:1 (fibres of this size are referred to as 'regulated fibres' by Doll and Peto[8] and are also known as 'optically equivalent' or 'optical fibres'). Many smaller asbestos fibres are detected when filters are analysed by electron microscopy, but in order to use the risk relationships derived from workplace exposure, it is necessary to express the results in terms of 'regulated fibres' which are used to express the results of workplace measurements. The reporting of asbestos fibres in terms of mass (ng/m^3) was abandoned some time ago as this measure cannot be directly related to health effects.

2.8 The lower limit of quantification of the measurements is determined by the volume of air samples, the size of filter used and the area of the filter analysed; the method used gave a limit of accuracy of about 0.1–1.0 fibres per litre (f/l). Most of the measurements gave results below the limit of quantification, but very small asbestos fibres were detected in many of the samples and some fibres of regulated size were found. It is possible to combine the results of several measurements made at the same time at a particular site to improve the limit of accuracy and obtain an approximate estimate of fibre concentration at that site. The fibre concentrations measured at a number of different sites can also be combined to produce an estimate of typical fibre concentrations found at those sites. The results of such an analysis gives an overall estimate of the concentration of asbestos fibres in the buildings in which measurements were made of 0.4 f/l.

The public health risk

2.9 Most of the information on the health effects of exposure to asbestos has been derived from studies of workers occupationally exposed to asbestos fibres at concentrations many times higher than those encountered by the general public. Estimates of the risk of low level exposure have to be based on extrapolation from occupational exposure levels, and the range of uncertainty in such estimates is large. The risk of mesothelioma is thought to increase rapidly with time since first exposure and it is therefore likely that children will be more at risk than adults from a similar exposure. Smoking and asbestos appear to act synergistically in causing lung cancer, and smokers exposed to asbestos have a much greater additional risk of contracting lung cancer than non-smokers similarly exposed.

2.10 Doll and Peto[8] reached the following conclusions on the risk to people exposed in buildings:

'The review of published studies by the Ontario Royal Commission (1984) and measurements made in British buildings on behalf of the Department of the Environment suggest that exposure to true asbestos fibres of regulated sizes within asbestos-containing buildings is seldom more than 0.0005 rf/ml [0.5 regulated fibres/litre] above background (as seen by optical microscopy). Exposure to this level for a working week in an office for 20 years in adult life or for 10 years or so at school, or to lower average levels for more prolonged times at home is calculated to produce a life-time risk of death of 1 in 100,000. If 20% of the population experience such exposure, this would imply that one death in a year was caused in the whole country.'

These estimates are based on the assumption of a linear exposure-risk relationship at low levels of exposure with no threshold or 'safe' level of exposure to asbestos.

2.11 This risk estimate is based on exposure to chrysotile asbestos and Doll and Peto point out that the risk could be appreciably greater for exposure to crocidolite and possibly amosite. There is a lack of quantitative data on the biological effects of exposure to crocidolite and amosite, but on the basis of the available information, the authors have estimated[13] that the risk of lung cancer and mesothelioma resulting from exposure to crocidolite might be 6 to 10 times greater than the risk from similar exposure to chrysotile. They believe that the increased risk from exposure to amosite will be less than this but point out that there is no good evidence on which to base a specific lower figure.

2.12 The results of monitoring indicate that the airborne asbestos fibres found in the buildings sampled were mainly chrysotile or amosite and only rarely crocidolite. There is a wide range of uncertainty in any calculation of risk, but in the most likely situation of mixed exposure, the risk would be possibly 2 or 3 times greater than that from exposure to chrysotile alone. This is still a very low level of risk.

2.13 The US National Research Council Committee on Non-occupational Health Risks of Asbestiform Fibres[9] calculated the following lifetime risks arising from continuous exposure to asbestos:

'If a person were to inhale air containing asbestos at an average of 0.0004 fibres/cm^3 [0.4 fibres/litres] throughout a 73 year lifetime, the committee's best estimate is that the lifetime risk of mesothelioma would be approximately nine in a million (range 0 to 350 per million, depending on assumptions regarding the relationship of dose to risk). The corresponding lifetime risk for lung cancer would be about 64 in a million for male smokers (range 0 to 290), 23 in a million for female smokers (range 0 to 110), and 6 and 3 in a million for male and female non-smokers, respectively'.

2.14 The risk arising from typical levels of exposure to asbestos in buildings found in the UK is very small, especially when compared to other common risks, such as road accidents or accidents in the house. Nevertheless, as there is no known threshold level for exposure to asbestos below which there is no risk, it is advisable to reduce exposure to the minimum that is reasonably practicable. In cases where there is potential for long periods of exposure, as in homes, or where children are involved, as in schools, particular efforts should be made to ensure that levels are as low as possible. In the general population, the risk of mesothelioma and lung cancer attributable to asbestos cannot be quantified reliably and are probably undetectably low. Cigarette smoking is the major cause of lung cancer in the general population. The risk of asbestosis is virtually zero.

Asbestos in drinking water

2.15 A survey of asbestos in drinking water, carried out for the DOE by the Water Research Centre, was published in 1984[14]. Results from 144 samples showed that levels of asbestos in water supplies in the UK were similar to or lower than levels in Europe or North America. However, there was evidence that asbestos-cement pipes can release fibres into the distribution system. This was most apparent where deposits of accumulated rusty chalk were disturbed near terminal hydrants from which samples were obtained. The report concluded that: 'The aggressiveness of the water, the length and age of the pipes and the occurrence of reversals of flow probably all contribute to the concentrations found'.

2.16 Two samples were taken from asbestos-cement water storage tanks. These were not enough to determine the contribution that such tanks might make to waterborne fibre levels. However, the report notes that 'much of the use has been in areas where the local water supply is hard and unlikely to corrode the tanks, and also for domestic hot-water storage rather than for storage of drinking water'. Unless an asbestos-cement storage tank is corroded internally there is no reason to consider its immediate replacement. If the tank has an asbestos-cement cover this could be corroded on its underside by condensation and, of so, should be replaced with a close fitting lid of another waterproof material.

2.17 The Department of Health's Committee on Medical Aspects of the Contamination of Air, Soil and Water was asked to advise on the implications for public health of the concentrations and forms of asbestos found in the survey, particularly in relation to the use of asbestos-cement pipes in drinking water distribution systems. The Committee concluded that:

'(a) The only potential risk from the presence of asbestos in drinking water which has been suggested as at all plausible, is that of certain forms of cancer;

(b) there is no convincing evidence from the substantial body of research findings relevant to this question which indicates that the concentrations and forms of asbestos in drinking water in the UK, including those derived from the use of asbestos-cement pipes according to current practice, represent a hazard to the health of the consumer; and

(c) if there is any carcinogenic risk to the consumer from exposure to asbestos in drinking water, it is of an extremely low order and is not detectable by the methods currently available.'

2.18 It must be stressed that water for drinking and cooking should – for reasons of hygiene unconnected with asbestos – be drawn only from the kitchen tap supplied directly from the main. It should not be drawn from any storage tanks in a roof space or from hot water systems fed via storage.

Table 1 **Results of airborne asbestos monitoring**

Site	Concentration of asbestos fibres per litre[a]	Type of asbestos
1. Factory. Sprayed asbestos on ceilings.	BLQ[b](1.0)	Chrysotile
2. Factory. Sprayed asbestos and asbestos-cement roof.	BLQ (0.2)	Chrysotile
3. Shopping centre. Sprayed asbestos on steelwork.	BLQ (0.1)	Amosite/chrysotile
4. College. Sprayed asbestos on ceilings.	BLQ (0.1)	Amosite
5. School. Sprayed asbestos on ceilings (damaged).	2.0	Amosite/chrysotile
6. School. Sprayed asbestos on ceilings.	BLQ (0.3)	Amosite
7. School. Sprayed asbestos on ceilings.	BLQ (0.2)	Amosite/chrysotile
8. School. Sprayed asbestos on ceilings.	BLQ (0.5)	Chrysotile
9. School. Sprayed asbestos on ceilings.	BLQ (0.4)	Chrysotile
10. School. Sprayed asbestos on ceilings.	BLQ (0.3)	Chrysotile
11. School. Sprayed asbestos on ceilings.	0.4	Chrysotile/amosite
12. School. Sprayed asbestos on ceilings. Removed 1 year previously.	0.7	Chrysotile/amosite
13. Laboratory. After removal of sprayed asbestos.	0.5	Crocidolite/amosite
14. Laboratory. Asbestos insulation on steelwork.	BLQ (0.8)	Amosite
15. Laboratory. Insulation material.	BLQ (0.3)	Chrysotile
16. Laboratory. Sprayed asbestos in cupboards.	BLQ (1.0)	Amosite/chrysotile
17. House. Textured plaster.	BLQ (0.1)	Chrysotile
18. House. Asbestos in warm-air heating.	0.3	Chrysotile
19. Flat. Asbestos insulation on steelwork.	0.4	Amosite/chrysotile
20. Flat. Asbestos insulation on steelwork.	0.7	Amosite/chrysotile
21. Flat. Suspected asbestos-cement panel in cupboard.	BLQ (0.3)	Chrysotile
22. Flat. Asbestos-cement panel in cupboard.	BLQ (0.5)[c]	
23. Flat. Asbestos-cement panel in cupboard.	BLQ (0.3)	Amosite
24. Flat. Suspected asbestos-cement panel in cupboard.	1.9	Amosite
25. Flat. Asbestos-cement boards in main room.	1.1[d]	Chrysotile
26. College		
(i) Before asbestos removal	< 0.2[e]	
(ii) 6 months after asbestos removal	0.8[e]	
(iii) 9 months after asbestos removal	0.4[e]	

[a] Results are reported in terms of regulated fibres, ie. length greater than 5 um diameter less than 3 um and aspect ratio greater than 3:1. (In the Control of Asbestos at Work Regulations 1987 and supporting Approved Code of Practice, control limits and action levels are expressed in terms of fibres per millilitre (f/ml); 1 fibre per litre = 0.001 f/ml).

[b] BLQ. Below Limit of Quantification. The concentration corresponding to 4 counted fibres is the limit of quantification. This concentration, which depends on the volume of air sampled, the size of filter used, and the area of filter analysed is given in brackets.

[c] No asbestos fibres detected.

[d] Samples were taken while the flat was being decorated.

[e] These concentrations support the view that stripping should only be undertaken if sealing or encapsulation are not applicable.

3 Asbestos materials used in buildings

3.1 This section describes asbestos materials that are or have been used in buildings. The installation of sprayed asbestos and thermal and acoustic insulation is now prohibited and asbestos insulating board is no longer manufactured in the UK. However, these materials may still be present in some buildings. Asbestos is still used in the manufacture of asbestos-cement and in materials such as mastics, sealants and protected metals. The tables in Annex 1 list the asbestos materials commonly used in buildings.

Sprayed coatings and lagging

3.2 The sprayed material applied in the UK was a mixture of hydrated asbestos-cement containing up to 85% asbestos fibre. Sprayed asbestos in buildings mainly contained amosite, but crocidolite may have been used in some installations. Although the import of crocidolite effectively stopped in 1970, existing stocks may have been used in later installations. Amosite was used for anit-condensation and acoustic control in buildings and for fire protection of structural steel. Chrysotile was used to a limited extent in sprayed asbestos until 1974, mixed with mineral wool and cementitious binders. It was also used as a coating over other sprayed asbestos fibre. Sprayed asbestos is unlikely to be found in ordinary houses. It is sometimes found in schools, for example on the ceilings of school swimming pools. It is a friable material and likely to release fibres, especially if disturbed during repair and maintenance work. As it ages the binding medium of sprayed asbestos may degrade with the consequent release of more fibres. The application of sprayed asbestos ceased in 1974 and the spraying of asbestos is now prohibited. The removal of asbestos insulation or asbestos coating is controlled by the Asbestos (Licensing) Regulations 1983 and the Control of Asbestos at Work Regulations 1987 (paragraph 1.10) which require that employers and self-employed persons, subject to certain exceptions, who carry out such work hold a licence granted by the HSE. Advice on complying with these regulations is provided in the booklet – 'A Guide to the Asbestos (Licensing) Regulations 1983' HS(R)19 Revised 1989. Practical advice and guidance on removing sprayed asbestos coating and lagging are set out in the Approved Code of Practice, 'Work with asbestos insulation, asbestos coating and asbestos insulating board.'[2]

3.3 Lagging is a term which covers a wide range of materials including pipe sections, slabs, rope, tape, corrugated asbestos paper, quilts, felts, blankets and plastered cement. Asbestos has also been used as a surface coating on felt and cork insulation. Asbestos lagging may have a protective covering of cloth, tape, paper or metal, or a surface coating of cement. The installation of asbestos thermal insulation is now prohibited. Crocidolite yarn and rope was used for lagging from the 1880s until the mid 1960s, although as cheaper chrysotile products became available it was generally restricted to uses where acid resistance was required. Crocidolite was used in insulation mattresses between the 1890s and the mid 1960s and in preformed thermal insulation between the mid 1920s and 1950. Amosite was used in preformed thermal insulation, pipes, slabs and moulded pipe fitting covers between the mid 1920s and late 1960s and for insulation mattresses between 1920 and the mid 1960s. Between the late 1950s and mid 1970s amosite asbestos was used to make reinforced calcium silicate high temperature insulation. From the middle 1960s onwards, man-made mineral fibre insulation materials, introduced in the early 1950s, took over most of the 'middle range' thermal insulation market.

3.4 The asbestos content of lagging depends on the type of material and can be high. Asbestos quilts, mattresses and blankets for example, may contain approximately 100% asbestos. A common form of pipe and boiler lagging consists of 85% magnesia (magnesium carbonate) and 15% asbestos, with an asbestos surface coat. Preformed thermal insulation materials made of magnesia, calcium silicate and diatomite were reinforced with some 10–15% amosite or a mixture of amosite and chrysotile. Low temperature insulation includes the felt or cork materials mentioned above. Asbestos based lagging has been widely used in public buildings, factories and hospitals. Magnesia insulation was particularly common on hospital pipework. Quilts were commonly used on steam boilers in industrial premises but were rarely used in houses or flats. Asbestos cord or rope was wound round pipework or insulation, sometimes covered with a hard cementitious compound which may itself contain asbestos. A small number of houses have had 'loose fill' asbestos loft insulation installed. This material was possibly obtained as waste from other work.

Asbestos has also been used for insulation between floors. The likelihood of fibre release from lagging depends on its composition, friability and state of repair, but it is particularly susceptible to damage or disturbance during maintenance work, or because of leaks from pipes or boilers.

Insulating boards

3.5 Insulating boards have a density of approximately 700 kg/m^3 and contain 16–40% asbestos mixed and hydrated Portland Cement or calcium silicate. They are frequently referred to as 'Asbestolux', a trade name. 'Shipboards' are a rigid composite of amosite asbestos, hydrated lime and silica with a density of 500–700 kg/m^3. Insulating boards were developed in the early 1950s to provide and economical, lightweight, fire resisting insulating material. The market for these boards expanded from 1950 until the middle 1970s when alternative fibre-reinforced boards became available. Asbestos insulating boards have not been made in the UK since 1980. Amosite was the normal type of asbestos used, although one manufacturer used approximately one third chrysotile to two thirds amosite in a lime silica board containing 27% asbestos. Insulating boards are found in all types of industrial, commercial, public and private buildings including houses, flats, schools and colleges. They are particularly common in 1960s and 1970s system built housing. They are mainly used to provide structural fire protection and heat resistance, acoustic insulation, partitioning, as a non-combustible core or lining for other products and, because of their resistance to moisture movement, as a general building board.

3.6 Asbestos insulating board ceiling tiles were made of board cut into squares and given bevelled edges, although some were also made with a perforated surface for acoustic absorption. The tiles, introduced in the early 1950s, were used in large numbers until the 1960s, when their use began to decline as substitutes became available. The tiles were widely used in schools, hospitals and shops. Insulating boards are semi-compressed and are therefore more likely to release fibre as a result of damage or abrasion. Work on asbestos insulating boards can give rise to high levels of asbestos fibre and advice on the required precautions is given in HSE Guidance Note EH37[15], and in the ACoP 'Work with asbestos insulation, asbestos coating and asbestos insulating board'.[2]

Ropes, yarns and cloth

3.7 The asbestos content of woven and spun materials approaches 100% and all three types of asbestos have been used in their manufacture.

Asbestos yarns, often reinforced with other yarns or filaments, were used in jointing and packing materials, gaskets and caulking for brickwork. Asbestos ropes have been widely used for thermal insulation of pipes and as a rotproof firestop where pipes pass through walls. Plaited asbestos tubing was commonly used as flexible insulation for electric wire and cable. Asbestos cloth was used in fire protective clothing such as overalls, gloves and aprons and in fire blankets and curtains, and was sometimes aluminized to reflect radiant heat. These articles are used in foundries, laboratories, kitchens or other places where there are very hot substances. The risk of fibre release depends on the structure of the material – a bonded gasket material is unlikely to release asbestos but an unbonded woven material could release fibres in use, especially if it is damaged or frayed.

Millboard, paper and paper products

3.8 These materials have an asbestos content approaching 100% and all three types of asbestos have been used in their manufacture. They have been used for insulation of electrical equipment and for thermal insulation, and asbestos paper has been used as a fire-proof facing on wood fibre board. They are not highly bonded and should not be used where they might be subject to abrasion and wear. They may be a hazard when handled, especially when large quantities are moved after storage.

Asbestos-cement products

3.9 Asbestos-cement products generally contain 10–15% of asbestos fibre bound in matrix of Portland Cement or autoclaved calcium silicate. All three types of asbestos have been used in the manufacture of asbestos-cement in the past. Crocidolite was used between 1950 and 1969 (imports of crocidolite ceased in 1970) and amosite from 1945 until at least 1976, but the majority of asbestos-cement is made with chrysotile fibre. Asbestos-cement may be compressed into flat or corrugated sheets or moulded into a wide range of components. The degree of compression of sheet materials is variable. Semi-compressed flat sheets have a density of about 1200 kg/m^3 with one smooth and one indented face. Fully compressed sheets have a density of about 1600 kg/m^3 and two smooth faces. Uncoated sheets are light grey in colour, but fully compressed sheets are available with a factory applied surface coating.

3.10 Corrugated sheets are largely used as roofing and wall cladding. Semi-compressed and fully compressed flat sheets are used largely as panelling or partitions, the degree of compression largely determining the strength of the material. The range of uses of sheet and moulded goods is described in

Annex 1. Some manufacturers have used cellulose pulp in their asbestos-cement mixes. The cellulose-asbestos-cement boards have densities (1200 mg/m^3 semi-compressed, 1500 kg/m^3 fully compressed) and strengths similar to ordinary asbestos-cement board, although these properties will vary with the cellulose content. The boards were used as flexible partition board or lining, with improved bending strength and lower thermal conductivity. Pigments were frequently added to give the boards a light tan colour. Asbestos wood was an asbestos-cement mixture which contained approximately 21% asbestos. It was intended for similar uses to fully compressed flat sheet, where a lighter board was required with good fire protection properties. Asbestos wood was also used on house doors where fire protection was required, but greater strength than asbestos insulating boards (paragraph 3.5) was needed. High density asbestos-cement sheet has been used in industry, workshops and laboratories for a variety of purposes.

3.11 The asbestos fibres in asbestos-cement are firmly bound in the cement matrix and will be released only if the material is mechanically damaged or deteriorates with age. Asbestos-cement, particularly when coated, is relatively resistant to light abrasion and impact. The lower fibre content and well bound matrix made asbestos-cement products are much less likely to generate dust during their normal operational life than sprayed coatings, lagging and insulating board. When used externally, asbestos-cement components will weather slowly, but the rate of fibre release is so low that it presents no significant hazard. It is possible that fibres may accumulate in certain areas such as gutters draining asbestos-cement roofs and care needs to be taken on cleaning out such gutters if dry. If asbestos-cement is used inside buildings, dust released by wear, maintenance or repair may accumulate. Asbestos-cement components should therefore not be used internally where they are likely to be subject to damage or abrasion, or where they will need to be cut or drilled or frequently disturbed after installation. Asbestos-cement may be painted on both sides to seal it completely, to prevent warping or cracking and to reduce surface deterioration (paragraph 3.16).

3.12 Although the Asbestos (Licensing) Regulations 1983 do not apply to work with asbestos-cement or asbestos insulating board having a density greater than 500 kg/m^3, they may nevertheless still be invoked if when working with asbestos-cement materials, other asbestos material is found to be present[5]. Asbestos dust can be released when working with asbestos-cement materials and HSE guidance on working with asbestos-cement[16, 47] should be consulted when planning such work. Small jobs carried out by a householder do not come within the scope of the health and safety at work legislation but nevertheless, the general safety precautions recommended by the HSE should be followed and these are summarised below.

3.13 When the work does not involve cutting, abrading or breaking the material and dust is not likely to be released, relatively simple precautions are sufficient.

Precautions should be taken to prevent any dust generation, for example by damping the material.

Any waste material should be collected and disposed of safely (paragraphs 6.2 – 6.9).

Any dust that is released should be removed with an industrial vacuum cleaner with a high efficiency (absolute) filter, commonly known as 'type H'. A domestic vacuum cleaner is not appropriate for use with asbestos and can create a significant airborne dust problem. Alternatively, a damp cloth can be used, sealed in a plastic bag while still damp, and then disposed of safely.

3.14 When asbestos-cement materials are cut or drilled dust may be released and additional precautions are needed.

- Only people engaged in the work should go into the working area.
- Protective clothing and an approved type of dust mask suitable for asbestos should be worn.
- A plastic sheet should be used to collect any dust.
- Hand tools rather than power tools should be used.
- Work should be done outside if possible and working overhead should be avoided.
- The material should be wetted to minimise dust release.
- Protective clothing should be cleaned or disposed of safely, and people engaged in the work should wash carefully.

3.15 Work which is liable to generate more dust (for instance work with power tools, building demolition, or removal of old disintegrating material) should be carried out professionally, and comply with the requirements set out in HSE Guidance.

3.16 After several years asbestos-cement used externally may become covered in lichens, algae or mosses. Such growth should have no noticeable effect on the strength, durability or lifetime of the structure although it may become visually unattractive. Unless the vegetative growth is removed or disturbed there will be no measurable release of asbestos fibres. In the first instance owners/occupiers should decide whether removal of the organic growth is needed, as cleaning operations may create unnecessary dangers of falling and risk of exposure to asbestos. If cleaning roofs is necessary, several methods are available depending on the circumstances. Wet methods should be used where possible in preference to dry methods. Water jetting produces problems of containment of slurry and spray, and precautions to prevent the spread of contamination will normally be necessary. Steps should also be taken to avoid asbestos-rich slurry entering buildings. The various methods will require the use of appropriate personal protective equipment including respiratory protective equipment. Any roof cleaning operation will require an assessment of asbestos exposure, control measures and environmental impact. Low pressure water and brushing can be used in conjunction with a biocide[17]. More detailed information is provided by the HSE[16] and this should be consulted before work begins. Asbestos-cement roofs can be fragile and at all times adequate precautions must be taken to prevent accidents.

Bitumen felts and coated metal

3.17 Some roofing felts, flashing tapes, damp-proof courses and other products contain asbestos fibre, sometimes in the form of asbestos paper, in a bitumen matrix. These materials are not likely to present a hazard during normal installation work or in use. It is possible that they could become brittle or beak up with age and they should then be removed carefully. Any adhering material should be removed manually (not by power grinding) and the waste material should be disposed of safely, not by burning. Roofing felts containing asbestos are no longer manufactured or imported into the UK.

3.18 Asbestos mixed with bitumen or bitumen reinforced with asbestos paper, has been used since the 1920s as a coating for corrugated steel sheet. The material has been used as a roofing and cladding for buildings such as warehouses and factories. The asbestos is firmly bound into the coating but can be released and dispersed of the bitumen burns in a fire. The protective coating should not be burned off scrap sheet.

Flooring materials

3.19 Asbestos has been added to the mix of certain PVC and thermoplastic floor tiles and sheet materials. Some types of PVC flooring have an asbestos paper backing. The type of flooring (for example 'cushion flooring') should be cut and laid carefully but should not be stuck down. Fibres bonded into flooring may be released as the material wears, but the rate of release is likely to very low except under conditions of very heavy wear. Worn flooring should preferably be covered, or removed. If removal is necessary, the material should be lifted carefully, and any loose dust dampened and collected; any residue sticking to the floor is best covered or 'skimmed', but may be carefully removed while damp. It must not be removed with a power sander. Waste material must not be burned but should be disposed of as described in paragraphs 6.2 – 6.9. Asbestos paper is no longer used as a backing for cushion floor made in the UK.

Textured coatings and paints

3.20 The supply and application of any paint or varnish containing asbestos for use at work is now prohibited by the Asbestos (Prohibitions) (Amendment) Regulations 1988. However, asbestos may still be found in some existing textured coating or paint, eg. 'Artex' and care should be taken when these are removed. These materials should not be sanded down or scraped off dry. If they must be removed, wettable materials can be removed by soaking with water containing a little detergent to soften them and then scraped off. Non-wettable materials containing asbestos should not be removed without taking the special precautions described in the ACoP, 'Work with asbestos insulation, asbestos coating and asbestos insulating board'[2].

Mastics, sealants, putties and adhesives

3.21 Small quantities of asbestos may be included in mastics, weatherproofing sealants, putties and adhesives to impart anti-slumping characteristics, to improve covering power and to prevent cracking and crazing. The fibres are bonded into the materials as sold as will not present any hazard during application. The only possible hazard is from subsequent sanding of hardened material. Some are thermoplastic and cannot be sanded but there could be occasion to sand-down others, for example vehicle underseals or tile adhesive. Sanding-down with power tools should be avoided. Appropriate precautions should be taken when working with these materials.

Reinforced plastics

3.22 Asbestos-reinforced PVC containing chrysotile asbestos has been used to make cladding and panels. Asbestos-reinforced plastics have also been used to make a variety of products including household items such as plastic handles and battery cases. Plastics products are unlikely to release fibres during use but cutting with high-speed power tools should be avoided.

4 Identification and assessment of existing asbestos installations

4.1 This section describes general procedures for identifying asbestos materials in buildings and for assessing the potential hazards from them. In general fibres are not released unless asbestos materials are disturbed or damaged and undisturbed materials in good condition present little risk. The tables in Annex 1 list the asbestos materials commonly used in buildings and indicate the conditions under which they are most likely to release fibre. The potential hazards from each separate installation of asbestos which is identified, for example a sprayed ceiling, a lagged boiler or a cupboard lined with insulating board, must be assessed to enable the correct remedial measure to be determined in each case. Selection of remedial measures is discussed in the next section.

Identification

4.2 Extensive use has been made of building materials containing asbestos, and unless the material is readily identifiable by for example, the use of an asbestos label (paragraph 7.18) the existence of substitute materials can easily be mistaken for the original. If they exist, a check of the original building plans may help to determine the presence of asbestos.

4.3 It is important to determine the type and amount of asbestos present in the building, the type of material in which it is contained, its use and its location. Some general points are given below and HSE Guidance Notes[2, 15, 16] give details of methods of sampling and analysis.
- Finding asbestos materials may require a diligent and determined search.
- Staff carrying out sampling must be suitably trained and experienced.
- Taking samples from asbestos materials will involve damaging the materials so fibres will be released into the air. Appropriate precautions must therefore be taken to avoid endangering the health of staff or contaminating buildings.
- Do not sample unnecessarily. The number of samples can be limited if each material concerned is treated as if it contained crocidolite asbestos on the basis of a single sample. Alternatively sampling may be avoided altogether if it is decided to treat all suspect materials as crocidolite asbestos.
- Arrangements should be made to repair damage, such as exposed edges, after sampling and to clean-up when sampling is completed.
- Waste material, cloths etc., should be disposed of as described in paragraphs 6.2 – 6.9.

Guidance on the number and size of samples needed for various types of material are given in Annex 2.

4.4 It will not usually be necessary to carry out an accurate quantitative analysis, and indication of the presence of asbestos and its types will be sufficient in most cases. Examination of the sample by polarized light microscopy will usually be adequate to give this information, but x-ray diffraction, or electron microscopy, may enable the type of asbestos to be identified when samples are unsuitable for analysis by light microscopy. Analysis of the samples should be performed by suitably trained and experienced analysts. Chemical tests may give confusing results and are not suitable for the identification of asbestos fibres.

Assessment

4.5 Once the presence of asbestos materials within a building has been established the potential for fibre release must be assessed so that appropriate remedial measures may be taken. Systems of assessment based on assigning weighted values to factors such as accessibility, friability, damage etc., are sometimes used. Trials of these risk rating systems have found considerable inconsistencies between individual survey staff and no correlation has been found between measured airborne asbestos concentrations and calculated ratings. Number based systems are not recommended for the selection of remedial measures for individual installations, but they may be useful in setting priorities for action (Section 5).

The potential for fibre release

4.6 The potential for fibre release from an asbestos material is determined by 3 main factors.
- The type of material and its properties, and the type of asbestos used in its manufacture.
- The integrity of the material, and the condition of any sealant or enclosure.

- The position of the material, which should take into account its accessibility and vulnerability to damage, and the use of the area or building where the material is installed.

The type of material

4.7 The composition of an unknown material and the type of asbestos present can be established by bulk sampling (paragraphs 4.3–4.4). Annex 1 and Section 3 give details on the asbestos materials used in buildings, including their asbestos content and notes on their useage. For the purposes of assessment, the types of asbestos material are listed in approximate order of ease of fibre release:

- sprayed coatings and laggings;
- insulating boards, insulating blocks and composite products;
- ropes, yarns and cloth;
- millboard, paper and paper products;
- asbestos-cement products;
- bitumen roofing felts, damp-proof courses, semi-rigid asbestos-bitumen products and asbestos-bitumen coated metals;
- asbestos-paper backed vinyl flooring;
- unbacked (homogenous) vinyl flooring and floor tiles;
- textured coatings and paints containing asbestos;
- mastics, sealants, putties and adhesives;
- asbestos reinforced PVC and plastics.

4.8 The hazard presented by these materials is related to their hardness or toughness and the ease with which fibres may be released. The ranking can only be approximate, but asbestos-cement for example, has a lower fibre content and is tougher than insulating boards or blocks; sprayed coatings and lagging have a higher fibre content and are softer still.

4.9 The performance of a material in use is to some extent determined by the type of asbestos from which it is manufactured. Crocidolite and amosite were commonly added to chrysotile in the manufacture of such products as asbestos-cement high pressure water pipes. The use of crocidolite and amosite tends to make products more friable with age than similar products made with chrysotile alone. Materials which contain crocidolite and amosite therefore tend to be less resistant to mild abrasion or damage, and generate more dust for equivalent work activity than do materials which contain only chrysotile.

The integrity of the material

4.10 For a material to be classified as being in good condition it must intact, not cracked or fractured and not bearing evidence of abrasion or fraying. Sprayed coatings should not be detached from the substrate. Paint, sealant or other coatings (where applied) should be adhering to the entire surface and any enclosure should be intact, including seals at corners and edges. In the area surrounding or beneath the asbestos material there should be no debris, or other evidence of dust release. Material should be classified as not in good condition if it is damaged, scratched or scraped, sealant has become detached or peeled, the enclosure is removed or broken, or tape seals removed and asbestos debris or dust is in the immediate area. Photographs of sprayed coatings, lagging and insulating board, illustrating varying degrees of damage, are presented in the centre pages of this booklet.

The position of the material

4.11 Readily accessible material is likely to be vulnerable to damage arising from vandalism, impact by vehicles, people, or objects and in certain circumstances, to damage arising from maintenance and repair work. Other sources of damage are vermin (rats, mice and birds) and water, which is particularly damaging to pipe and boiler lagging and sprayed asbestos. If an anticipated future use of a building would make previously undisturbed asbestos installations accessible and therefore vulnerable to damage, the material concerned should be reassessed. The number of persons exposed and the extent of their exposure will be determined by the building use and occupancy. This will strongly influence the priorities for action discussed in Section 5. In schools, the location and accessibility of the material to children and the pattern of children's behaviour, should also be taken into account when deciding on a course of action.

Measurement of airborne asbestos fibre concentrations.

4.12 The purpose of any measurements should always be carefully considered before they are commissioned. If the condition of the asbestos material is bad enough to require remedial action anyway, measurement of airborne concentrations beforehand may not have practicable benefits. Measurement may provide useful additional information in doubtful cases, or may reassure occupants and members of the public when the asbestos has been refurbished, or appears to be in good condition.

4.13 Even damaged material will often not release many fibres unless it is being disturbed, so sampling should take place while activity in the area is representative of that occuring in normal occupancy or use. An appropriately more energetic disturbance procedure may be used to stimulate a 'worst case'.

The measurement of airborne dust concentrations and air sampling procedures should follow those described in HSE guidance EH 10[18] and MDHS 39/3[19] for assessment in relation to the 'clearance indicator'. This includes taking sample volumes of at least 480 litres on membrane filters. Under dirty conditions, it may be necessary to use smaller sample volumes to obtain countable samples. A number of the smaller-volume samples should then be taken, and the counts and volumes combined to give evaluations based on 480 litre volumes (minimum) as described in MDHS 39/3.

4.14 Analysis of the sample consists of counting the fibres on representative filter areas by either phase contrast optical microscopy (PCOM) or electron microscopy. The PCOM method is described in MDHS 39/3. It gives results which can be nominally compared with occupational exposures and epidemiology, but asbestos fibres cannot positively be distinguished from non-asbestos fibres by this method. Even in buildings containing asbestos materials, the majority of fibres are likely to be non-asbestos such as gypsum or cellulose. Supplementary techniques, such as polarised light microscopy, can be used to eliminate the larger non-asbestos fibres from the count. The limit of quantification of the method (as described in MDHS 39/3) is about 0.01 fibres/ml, and a concentration result above this indicates that either remedial action or measurement of the asbestos concentration by electron microscopy, is required. However, 0.01 asbestos fibres/ml is not an environmentally acceptable level for continuous public exposure, and the presence of asbestos fibres at lower concentrations must be taken into account with other factors in the assessment described in paragraphs 4.5 – 4.11. Laboratories should only be used for PCOM measurement if they have been accredited for the technique by the National Measurement Accreditation Service (NAMAS) (paragraph 4.16). Sampling should be carried out by competent and trained personnel. Although NAMAS does not yet accredit for sampling, laboratories accredited for PCOM can usually provide a (non-accredited) sampling service.

4.15 Electron microscopy is of two types, scanning (SEM) or transmission (TEM). In both cases, the instrument must have an X-ray analysis facility to determine whether or not the fibres are asbestos. SEM is best regarded as a refined optical method, capable of determining concentrations of fibres in the size range to which PCOM applies, but with the advantage that the analysis permits elimination of non-asbestos fibres from the count. Under normal conditions of use, the diameter of the smallest fibre which can be analysed by SEM is about 0.1 to 0.2 microns or micrometres (ie. one millionth of a metre), about the limit of fibre visability of PCOM. TEM permits counting and analysis of all fibre sizes, and the electron diffraction facility permits indentification in the rare cases where energy-dispersive X-ray analysis gives an uncertain result. The concentrations by TEM of asbestos fibres in buildings known to contain asbestos is generally < 0.03 fibres/ml (all fibre sizes) and <0.001 fibres/ml (length >5 microns or micrometers, all diameters).[40, 41]

Quality control and selection of laboratories

4.16 To comply with Regulation 15 of the Control of Asbestos at Work Regulations 1987 employers should ensure that laboratories which they use for sample analysis have the necessary facilities, expertise and quality control procedures to provide accurate results. Good quality control procedures are essential because of the large variations in results which can otherwise occur within and between laboratories with all fibre-counting methods. In most cases the only feasible way of doing this will be to choose a laboratory accredited by NAMAS for airborne asbestos measurement by methods set out in HSE guidance. Among other things an accredited laboratory will have been independently assessed. NAMAS is the National Measurement Accreditation Service, the acronym NAMAS has superceded the term 'NATLAS' used in the ACoPs, though laboratories may still refer to themselves as NATLAS and NAMAS accredited. If an accredited laboratory is not used, the person in charge of commissioning the work should be satisfied that the laboratory chosen works to an equivalently high standard. A list of the requirements that non-NAMAS accredited laboratories should satisfy are set out in HSE Guidance[18].

5 Remedial measures, surveys and priorities

5.1 The previous section described general procedures for identification of asbestos installations and assessing the potential hazards from them. Once this potential has been assessed for individual asbestos installations, appropriate remedial measures may be determined for each one. Individual assessments may be part of a larger survey, covering a range of buildings or building types. When the results of such a survey are compiled it is necessary to allocate an order of priority to any remedial work which is to be carried out. This section discusses remedial measures, surveys and priorities. It is not possible to give detailed instruction for all the many different uses of asbestos materials which may be found in buildings, but the guiding principles are:

asbestos materials which are sound, undamaged and not releasing dust should not be disturbed;

the release of asbestos dust should be avoided as far as possible;

the concentration of airborne asbestos in occupied areas should be reduced to the lowest reasonably practicable level.

5.2 The four charts in Annex 3, together with their explanatory notes, set out a procedure for the assessment of the potential of asbestos materials to release fibres, and provide guidance on the selection of remedial measures. The diagrams take account of the factors described in paragraphs 4.5 – 4.11 and use a systematic 'decision tree' approach, but they are not intended to give rigid rules for every case. Chart 1 summarises the decisions to be made when asbestos materials are first identified and leads to a decision to manage the materials in place, or directs attention to Charts 2 to 4, which give directions for sprayed asbestos and lagging, asbestos insulating board and asbestos-cement, the major uses of asbestos in buildings. Each diagram starts at the top of the page and moves downwards. Diamond shaped boxes represent questions which must be answered yes or no, rectangles represent the recommended remedial action. Separate notes on the other asbestos materials, which generally have small usage or present very little risk of fibre release in normal use, are given in Annex 4.

Remedial measures and management

5.3 The remedial measures available are:
- leave the material in place without sealing and introduce a management system;
- leave the material in place, effectively seal (eg. encapsulate or protect by mechanical means – board materials etc.), and introduce a management system;
- remove and dispose of the asbestos.

These actions vary in detail depending upon the type of material, its condition, location and accessibility, but are outlined in general terms below.

Management

5.4 It is not normally necessary to seal, enclose or remove asbestos materials which are sound, undamaged and not releasing dust. These should be left in place and a system of management introduced, which will require some or all of the following steps to be taken.

- The presence of an asbestos material should be noted on plans or other records, and updated as necessary. It may be appropriate to set up a register of the location of asbestos materials in buildings.

- Building owners should make known the existence or suspected existence of asbestos to any contractors working on their premises, preferably at the tendering stage.

- Maintenance and other workers, and other people who may be affected should be notified. In housing, occupants, especially those in public sector rented accommodation, should be made aware of the location of any asbestos materials and advised of appropriate precautions.

- Asbestos materials likely to be disturbed by maintenance or other workers should be labelled clearly. The 'a' logo (Annex 5), or similar, is recommended. It should not normally be necessary to attach labels to materials in housing, classrooms or repeated installations in one location, for example ceiling tiles.

- A note should be made of which types of work on the material will require licences, etc., and adequate instructions must be given to the workforce.

- The installation must be reinspected periodically to ensure that the condition of the material has not changed. The period between inspections will depend on the type of material and it may be possible to combine reinspection with other maintenance work. Friable and vulnerable materials should be inspected more frequently.

- Minor repairs should be carried out if necessary.
- If an anticipated future use will make undisturbed asbestos vulnerable to damage, management alone may not be sufficient and, as further action may be necessary, a reassessment should be carried out.
- It may be helpful to record the locations of non-asbestos materials which may be confused with asbestos materials.

Where management over an extended period is unlikely to be cost effective, further action, including removal, should be considered.

Sealing and repair

5.5 Sealing (or encapsulation) requires the application of some form of coating, whether paint, polymeric, bituminous or cement based. The sealing system chosen will depend on the nature of the asbestos material, the degree of damage, the protection required and surface flammability requirements. The sealing coat must adhere firmly and the integrity of the asbestos material must be sufficient to carry the sealing coat. Asbestos materials must be firmly attached to the substrate. Where asbestos insulation has been used to provide fire protection, the fire hazard must not be increased by the use of combustible sealants. The sealed materials must meet the standard for spread of flame specified in building regulations (as defined in British Standard BS476). Normal paints may not achieve this standard and specially formulated sealants are available.[20] Asbestos material which is damaged and is to be left in place, with or without sealing, must be repaired. This means patching, filling cracks, etc. and repair to enclosures, sealant or encapsulation where these treatments have already been carried out. Adequate precautions should be taken.[2, 15, 16]

Enclosure

5.6 Asbestos materials may also be enclosed with sheet material sealed at corners and edges. However resistance to fire spread must be maintained. The enclosed area between the covering and the asbestos material should be sealed and adequate cavity fire barriers constructed.

Removal

5.7 When it is not possible to seal an asbestos material effectively and it is likely to release dust, it may be decided to remove it completely. Removal may be the most cost effective solution in situations where the asbestos material may be disturbed frequently, for example during maintenance. However, in the short term, removal may lead to higher dust levels than sealing in situ (see Table 1), and as these elevated levels may persist for several months, appropriate precautions must be taken. Depending on circumstances the asbestos removal may be complete or just restricted to a smaller vulnerable area. Temporary repair, sealing or enclosure, may be required to render asbestos material safe pending removal. When asbestos fire protection material is removed it must be immediately replaced with materials having at least an equivalent fire performance. The Approved Code of Practice, 'Work with asbestos insulation, asbestos coating and asbestos insulating board'[2] provides practical guidance on the removal of asbestos. Removal of sprayed asbestos, lagging and low density insulating board, falls within the scope of the Asbestos (Licensing) Regulations[5]. The work should generally be carried out by a contractor licensed by the HSE and will be governed by the Control of Asbestos at Work Regulations 1987 and its associated Approved Code of Practice.[6]

5.8 After asbestos has been removed, the enclosure must be thoroughly cleaned and inspected (see HSE Guidance Note EH 51),[43] and the airborne concentration of asbestos fibres measured by the clearance indicator method in MDHS 39/3[19], to ensure that the concentration is below the level of quantification by PCOM (paragraphs 4.13 – 4.15). It is recommended that the laboratory used is NAMAS accredited (paragraph 4.16) and independent of the asbestos removal contractor.

Surveys

5.9 When large numbers of buildings are involved or where asbestos is extensively used in a single large building such as a block of flats, it may be necessary to carry out a survey for asbestos materials. This will involve inspection of individual buildings and indentification, assessment and selection of appropriate remedial measures for each separate asbestos installation within the building. A complete survey intended to cover every use of asbestos in large numbers of buildings will be expensive and unrealistic, but it is possible to reduce the cost by examining a representative proportion of buildings in detail and the remainder in less detail. For example, a survey of houses or flats in which the same fundamental design is repeated, will only need to examine approximately ten per cent of units of a particular type; but at least one unit from each design type must be examined. The results of a survey will enable priorities for remedial measures to be decided and a long term programme of work or management to be devised. Specific measures in individual dwellings which may be necessary due to variations in the general design, can be decided when each dwelling is visited as part of a planned programme of remedial works.

5.10   Appropriate survey strategies might be to

give priority to particular types of building (schools, blocks of flats, etc.) where large numbers of people, particularly children, may be exposed;

concentrate on building types in which it is suspected that there is a large scale, repeated, or consistent use of asbestos materials such as sprayed insulation, insulation board, or asbestos-cement;

concentrate on sampling more 'friable' materials in more highly utilized areas – for example determining the extent of sprayed asbestos coating or insulating board in communal areas of flats, or pipe and boiler lagging vulnerable to damage arising from frequent disturbance during maintenance.

The choice of strategy will depend on the extent of available information on the use of asbestos materials and on the number of buildings and range of building types to be considered. Buildings such as depots, libraries, swimming baths and hospitals which were built on an individual basis, will need to be examined separately.

5.11   Surveys should be conducted by specialist teams. It is desirable to include people with a knowledge of building construction and design, trained and experienced in techniques of sampling and recognizing asbestos material in situ. It is essential that all observations are recorded at the time they are made and a model survey report form is illustrated in Annex 6. Details of building type, construction, number of rooms, usage and occupancy should be noted. Each separate location of suspect material should be recorded, together with details of type (sprayed coating or lagging, insulating board or tile, asbestos-cement products, or 'other') and function (fire prevention, lagging, ceiling tile, wall partition etc.). It may be helpful in devising a management programme to note whether the asbestos material is used internally or externally, and a photograph of the installation may prove a useful aide-memoire for future reference. If a sample of the material is taken for analysis (paragraphs 4.2–4.4), this should be noted and the results of analysis recorded when they are available.

5.12   Where a large number of buildings of identical design are to be surveyed, it will only be necessary to take samples from suspect materials in the first few buildings visited. When the results of analysis are available and the suspect materials are identified, similar materials in identical locations in other buildings of the same type, may be identified by their appearance.

Priorities

5.13   Having identified asbestos materials in buildings and assessed the potential hazard of each installation, it will be necessary to decide on what action is necessary and in what order the work should be undertaken, particularly where large numbers of installations and large numbers of buildings are involved. Common sense dictates that remedial measures should be carried out on a basis of 'worst case' first. The ranking of installations of asbestos materials in order of urgency may be assisted by a priority rating system. To be effective any such system should rank sites in order of potential fibre release, so that there is a definite order of priority, and should identify those sites requiring immediate attention and those where the potential risk is very low. Such a system will necessarily involve a large amount of data storage and handling and numerical computation. This may be made much easier by the use of a computer.

5.14   Where the range of priorities is not very great it may prove cost effective to organize remedial works on a neighbourhood basis, undertaking higher and slightly lower priority work at the same time, rather than carrying out work on a discontinuous site-by-site basis strictly according to the priority rating. It may be more efficient in some cases to reduce the priority of 'high priority' removal work by temporary sealing or encapsulation, enabling the work of removal to be undertaken at a later date.

5.15   It will generally be more cost effective to undertake work on asbestos materials as part of a programme of general refurbishment or maintenance than to undertake such work separately. Any programme of work should therefore take account of other planned work and factors such as the availability of suitable contractors and temporary re-accommodation of persons whilst asbestos work is carried out. It is important that there should be adequate liaison with tenants, tenants associations, contractors, consultants and employees.

5.16   The Department of Education and Science[21] outlines a suggested approach for the determination of maintenance priorities in schools and colleges, and allocates first priority to work needed to ensure the health and safety of building occupants and users. It suggests that where there is no very high priority, remedial works should be carried out as part of normal maintenance programmes. Wherever possible, removal work in schools and colleges should be carried out during the holidays.

6 Demolition, disposal and fires

Demolition, including modifications and refurbishment

6.1 Demolition of buildings, the fabric of which contain asbestos materials, can release very high concentrations of asbestos dust into the atmosphere unless appropriate precautions are taken beforehand. It will be necessary to undertake a thorough survey of the building before demolition commences to discover the location of asbestos materials and other hazardous substances, although the full extent of the use of asbestos may not become apparent until work is under way. Asbestos insulation and coating should be removed wherever possible before any other demolition work commences, the removal of which should be performed only by contractors licensed under the Asbestos (Licensing) Regulations.[5] Removal involving all forms of asbestos (including asbestos-cement, asbestos insulating board and other non-insulation uses) is subject to relevant health and safety legislation which lays down the precautions to be taken to control and minimise release of asbestos dust[2]. It is a condition of most licences issued under the Asbestos (Licensing) Regulations that notice is given to the enforcing authority before the work begins. Advice on demolition of buildings containing asbestos materials is published by the HSE[22]. In certain cases the local authority must be informed of the intention to demolish a building, and may impose conditions on the demolition work (paragraph 1.13).

Disposal of asbestos waste

6.2 Any waste containing asbestos is classed as a controlled waste under the Control of Pollution Act 1974 (Chaper 40). Its disposal can be only to sites licensed under the Collection and Disposal of Waste Regulations 1988 (paragraph 1.10). In England in all areas other than Greater London and the other metropolitan areas, the waste disposal authorities are the county councils. In Greater London the London Waste Regulation Authority is the body responsible for controlling the disposal of asbestos wastes, and in Merseyside and Greater Manchester it is the Merseyside Waste Disposal Authority and the Greater Manchester Waste Disposal Authority respectively. In Tyne and Wear, West Yorkshire, South Yorkshire and the West Midlands, the metropolitan district councils are the waste disposal authorities. In Wales, Scotland and Northern Ireland, the waste disposal authorities are the district (or islands) councils.

6.3 Wastes[23] which are particularly dangerous are classified as 'special wastes' under the Special Waste Regulations (paragraph 1.10). These regulations are additional to the regulations covering the collection and disposal of controlled wastes and provide for control over the movement of waste from its point of arising to its final disposal using a consignment note system[24]. (A consultation paper reviewing these regulations was issued in January 1990). Special wastes include:

any measurable quantity of crocidolite or;

any other form of asbestos containing at least 1% by weight of free asbestos fibres or dust.

Sprayed coatings, lagging, broken insulating board which may be releasing dust, also other asbestos wastes containing crocidolite, are classified as special wastes. Bonded asbestos products such as bitumen felt and coated metals, PVC floor tiles, some decorative coatings, eg. 'Artex', and asbestos reinforced plastics, provided they do not contain crocidolite, are controlled wastes. Asbestos-cement is a controlled waste unless it has been broken into small pieces, pulverized, or in a form releasing substantial amounts of dust or contains crocidolite; in that case it would be a special waste.

6.4 Suitable receptacles must be provided for transport for special wastes, which are subject to The Dangerous Substances (Conveyance by Road in Road Tankers and Tank Containers) Regulations 1981, (Statutory Instrument 1981 No. 1059) and the Dangerous Substances (Carriage by Road in Packages) Regulations 1986 (Statutory Instrument 1986 No. 1951) and supporting Codes of Practice. Tank containers with capacity greater than 3 m^3 must be labelled to accord with the Dangerous Substances (Conveyance by Road in Road Tankers etc.) Regulations. Tank containers having a capacity less than 3 m^3 and most packages which contain asbestos waste must be labelled in accordance with the Classification, Packaging and Labelling of Dangerous Substances Regulations 1984 (Statutory Instrument 1984 No. 1244) (the CPL Regulations) with for example, the label illustrated in Annex 7, including a short statement on the nature of the dangers to which the substances may give rise. Receptacles (for example impervious sacks or drums) containing less than 25 kg of asbestos waste each must be labelled in a similar manner to large containers, but need not be accompanied by the statement. It should be noted that the label requires the name, address and telephone number of the *supplier* of the material. As specialist contractors often remove and transport asbestos waste, the removal contractor should be the named producer of the waste. The consignment notes required by the Special Wastes Regulations require the *source* of the material to be named. This is generally taken to be the owner of the building or area from which wastes are removed.

6.5 Fibrous waste or small off-cuts which should be free of sharp edges may be packaged and conveyed in double plastic sacks of suitable strength. The inner sack should not be overfilled and each sack should be capable of being securely tied or sealed. It is important that the label referred to above is clearly visible. If an opaque outer sack is used, it must be labelled; if a clear outer sack is used the label may be on either the inner or outer sack, provided that it is clearly visible.

6.6 Controlled wastes which are not special wastes would not normally be subject to the CPL Regulations, or the Dangerous Substances (Transport) Regulations. However, as a consequence of the Control of Pollution (Amendment) Act 1989, the registration of all carriers of controlled waste are in prospect. Small pieces of asbestos containing material are best placed in suitable duty plastic sacks. Large pieces of asbestos-cement sheet and pipe and other bonded material should not be broken or cut. Aged material showing signs of surface deterioration should be wrapped and carefully transferred to a suitable container, for example an enclosed skip. Enclosed or covered skips, unless used as tank containers, should be labelled with the 'a' logo (Annex 5).

6.7 Householders, who have asbestos waste for disposal should contact their local authority waste disposal department (paragraph 6.2) if they have asbestos waste for disposal. Small quantities of asbestos waste arising in the home should be dampened to prevent dust or fibres escaping and then sealed in a strong plastic bag marked 'ASBESTOS'. Large pieces of asbestos-cement sheet or pipe should not be broken up and, unless showing signs of surface degradation, need not be sealed in bags. Where practicable a label should be attached. The asbestos waste should be kept separate from other waste.

6.8 Waste disposal authorities have a number of options for dealing with household asbestos waste, depending upon local circumstances. Compaction or pulverization of asbestos waste should be avoided to prevent the release of respirable fibres or dust.

- Collection authorities may collect the waste and take it to an appropriate collection point before final disposal or direct to a licensed site.
- Waste disposal authorities may advise householders to take the waste to a collection point or nominated disposal site.

If waste is taken to a collection point, the authority should ensure that suitable facilities and adequate control and supervision are available.

6.9 Most asbestos waste is disposed of by landfill on licensed sites. The procedures for storage, transport with disposal of waste and the maintenance of landfill sites are covered in the 'Code of Practice for the Disposal of Asbestos Waste' from the Institute of Wastes Management.[48] This Code was drawn up by a Committee incorporating members from industry and the relevant Government Departments. The Code expands upon and brings up-to-date the information contained in Waste Management Paper (WMP 18) from the DoE[26]. It is important that all asbestos waste is handled and disposed of properly. Further advice may be obtained from the HSE or a local waste disposal authority office.

Asbestos in fires

6.10 Asbestos fibres change their mineral structure after prolonged heating, losing both their fibrous nature and mechanical strength. However, DoE has carried out research into the fate of asbestos in fires, which has shown that typically only the outer layers of asbestos materials are changed, and therefore potentially hazardous fibres remain within fragments of fibre debris. Thus, following a fire involving asbestos the precautions outlined below should be followed.

6.11 Experience of real fires, such as that at the MOD Warehouse at Donnington, has shown that some debris containing asbestos will remain at the site of the fire and some may be dispersed over a wide area. Severe fires and explosions in buildings which are clad or roofed with asbestos-bitumen may cause the coating to burn off from the metal, generating a grey paper-like ash which may contain changed and unchanged asbestos fibres. Asbestos-cement can explode when involved in fires, spreading changed and unchanged asbestos-cement debris over a wide area.

6.12 DoE research, on asbestos fibres which have been subjected to heat, has shown that identification of fibre types can be difficult, and that optical analysis of such fibres must take account of the changes in the microscopic characteristics.[29] However, further research on the examination of debris from actual fires has shown that identification of fibre types within a material should not normally present problems. Relatively unaffected samples will always be available for analysis, and fire damage seldom penetrates the outer layer of asbestos materials. In such samples, identification of fibre types will be possible provided that a sufficient number of fibres are viewed. A random selection of blackened debris in order to assess the presence of asbestos is not recommended. Wherever feasible, sampling should concentrate on products most visually similar to unaffected asbestos materials.

6.13 The binding matrix of asbestos materials can become weakened during fires and to minimise any subsequent release of fibres from solid debris or ash, for example by trampling it underfoot, it should be dampened down gently and collected carefully with the minimum physical disturbance as soon as possible. The appropriate precautions to protect the person should be taken and the ash and asbestos debris should be sealed into strong plastic bags (paragraph 6.5), labelled as indicated in Annex 7, prior to disposal. Where it is known that a building contains asbestos materials, appropriate precautions should be taken before debris remaining at the site of a fire is disturbed.

7 Appliances and equipment containing asbestos

7.1 Asbestos has had a variety of uses in the home, in schools and colleges. In addition to building construction it has been used in some domestic appliances, household goods and materials for DIY work. It may have been used in fixtures and fittings such as fireplace surrounds, the lagging of older central heating systems, asbestos-cement flue pipes and asbestos millboard covers for fuse boxes. In the majority of applications, the likelihood of fibre release from undamaged asbestos-containing products in situ is very small. The background level of airborne asbestos fibre found in houses is normally extremely low and poses negligible risk to the occupants. However, it is particularly important to minimize the release of asbestos dust in the home as people may be exposed for long periods of time.

7.2 The mere presence of asbestos materials does not constitute a hazard, and removing undamaged material may release more dust than would leaving it in place. It is only when asbestos materials are in a damaged or friable condition, or during installation, modification, removal or demolition, that there is likely to be significant fibre release.

Domestic appliances

7.3 Asbestos has been used in domestic appliances for its thermal or electrical insulating properties, as a friction material and in gaskets and seals. UK manufacturers no longer use asbestos in hair driers, fan heaters, irons, toasters, washing machines, tumble driers, spin driers, dish washers, refrigerators or freezers, except for small amounts in some appliances in the form of gaskets or braking pads, which are usually sealed within the appliance and are unlikely to release free fibres into the atmosphere.

7.4 Asbestos may still be found in some older types of cooker as oven linings (asbestos board), sealing between metal panels in the oven (asbestos fire cement) or oven door seals (asbestos rope). Asbestos rope has also been used as a seal on the lids of some solid fuel stoves. Any dust deposited in cookers should be wiped up with a damp cloth which should then be disposed of safely. The use of asbestos materials in seals is being phased out in domestic appliances, although some rubberized asbestos gaskets, which are not a hazard to health are still used. Asbestos-free sealants are used in most new appliances, although asbestos rope is still available and may be used by some fitters or service engineers. Asbestos-free materials can be specified when buying replacement parts or requesting service work. The installation of asbestos materials for thermal insulation is prohibited by the Asbestos (Prohibitions) Regulations which came into effect on 1 January 1986.

Household goods

7.5 Asbestos textiles are no longer used in oven gloves manufactured in the UK, and a wide range of substitute materials are available. Simmering mats made from asbestos millboard or asbestos paper may still be available and are a potential source of exposure if they flake with ageing. Mats showing signs of deterioration should be disposed of safely. Substitute mats made of metal are available. Older types of ironing boards may have iron rests made of asbestos milling board or paper. These are liable to wear with age and should be replaced if they show signs of damage or wear. In recent years, iron rests have been made of asbestos-cement, which is much more hard wearing than millboard or paper, and contains a smaller percentage of asbestos; it is resistant to abrasion, and the asbestos fibres are bonded into the material. Unless damaged by impact it is unlikely to present a hazard to the consumer. Asbesto-free iron rests are now widely available.

Asbestos fire blankets

7.6 Asbestos fire blankets are commonly found in laboratories, schools and catering establishments. The blankets are normally stored in containers and any risk of fibre release during storage in these conditions is negligible. However, it may be necessary to inspect blankets to ensure that they are still serviceable. Ideally, to avoid the risk of fibre release, inspections should be carried out without removing blankets from their containers. In addition, the following checks should be made on blankets held in cylindrical containers to ensure that they can be quickly and safely withdrawn in an emergency:
- wearing a half mask respirator approved by the HSE, try to move the blanket to ensure that it is not jammed, or so loose that it falls out;

- check any locking pin or flap to see that it is free;
- inspect the base of the container to see that there are no sharp edges that could cause injury, or damage the blanket.

If the blanket and container are obviously in satisfactory condition, they should be left in place ready for use. If their condition cannot be assessed, the container should be taken to a safe place in the open air. If the container cannot be moved without removing the blanket and before is removed in the open the inspector should put on a half mask respirator (approved by HSE). The inspector may then make appropriate checks on the container, remove the blanket, open it, and assess its condition. A blanket which is undamaged and uncontaminated (for example, with cooking oil or flammable chemicals) may be considered satisfactory, but if in doubt the blanket should be sealed in a polythene bag and the local fire prevention officer consulted. A blanket in satisfactory condition should be sealed in a clear, readily tearable polythene bag, which is labelled 'Asbestos' or with the 'a' logo label in Annex 5. The bag may then be stored in a container of appropriate size, ready for use. The container should not be sealed. As blankets are unlikely to deteriorate in storage, subsequent inspections should be confined to checking that the blanket is present, and that blanket, bag and seal are not obviously damaged – the blanket should not be removed from the bag. Following use in a fire, blankets should be placed in polythene bags and reinspected. Blankets and containers in unsatisfactory condition should be disposed of as asbestos waste (paragraphs 6.2 – 6.9). It may be useful to record the site and condition of fire blankets following inspection. Asbestos-free fire blankets are noted in paragraph 8.11

Asbestos in heating systems

7.7 Asbestos has been used for thermal and electrical insulation in a number of different types of heating system including catalytic gas heaters, electric storage heaters and gas warm air heaters. Various types of asbestos material have been used at different times and some models contain asbestos, other not. In order to identify heaters which contain asbestos, information is needed on the exact model and when it was made.

7.8 Some types of gas heaters, fuelled by liquified petroleum gas, contain loosely compressed asbestos fibre panels. These catalytic gas heaters 'burn without a flame' – the gas is oxidized as it passes over the asbestos panel, which contain a platinum catalyst. The panels may contain up to 0.5 kg of chrysotile asbestos. These may crumble as they age and asbestos fibres can be released into the air. The deterioration may be apparent on visual inspection as bald patches within the panel, or as a dust deposit on the surface below it. The fibre in earlier types of panel was only held in place by wire meshes. Later types were covered by another, woven material (probably man made mineral fibre) but this does not necessarily seal the asbestos in. The Department of Trade issued a public warning about the heaters and sent detailed advice to environmental health departments in 1983[27]. A one year prohibition order on their sale was introduced in November 1983[28], and a permanent ban came into effect on 21 November 1984[28A]. Members of the public in any doubt about the condition of such a heater should contact their local authority environmental health department for advice.

7.9 Asbestos materials have been used in some gas-fired warm air heating systems as insulation around the fan chamber or heat exchanger unit. There is unlikely to be a signficant risk of fibre release arising from the normal use of these heaters, but asbestos could be released if the insulation is damaged during installation, or during maintenance or repair. It is also possible that some types of insulation would degrade with time if subjected to heat. British Gas regional offices may be able to advise whether a particular warm air heating system contain asbestos material, and whether it is advisable to inspect the system to determine the condition of the insulation and see if there are signs of deterioration or dust release. If asbestos insulation is clearly deteriorated it may be replaced with a non-asbestos material with equivalent performance, approved by the manufacturer. Alternatively it may be more economic or practicable to replace the entire unit, especially if it is approaching the end of its useful life.

7.10 Asbestos insulating board and other asbestos based materials have sometimes been used for fire protection in the construction of cupboards and heater cupboard doors, including airing cupboards. Insulation used in such circumstances should be checked and if necessary sealed as described in paragraph 5.5, or it may be replaced with an asbestos-free substitute material with equivalent fire resistance. Asbestos materials may also have been used to line warm air distribution ducts. Measurements suggest that there is unlikely to be significant fibre release from undisturbed material in good condition, even where air is blown across them. However, ducting may sustain damage from maintenance of other building services, or from leaking pipes etc., and maintenance staff should be warned to take care to avoid damaging it.

7.11 Asbestos lagging consisting of an asbestos layer sealed in an asbestos/plaster coat has been used in the past to lag boilers and pipework in some older coal-fired domestic hot water/central heating systems. Unless the material is damaged during maintenance or repair work by leaking water, it is not likely to release fibre. If these older central heating systems are removed, the work will be subject to the Asbestos (Licensing) Regulations.[5] Boilers should, where possible, be removed with the lagging intact and disposed of as asbestos waste.

Electric 'warm air' and storage heaters

7.12 Asbestos materials have been used to provide insulation in older models of electric storage heaters and larger warm-air heating systems. Advice has been issued by the Electricity Council,[30] (now the Electricity Association) including a list of the types of heater. Three types of appliances were involved:

- storage heaters which operate primarily by radiant heat;
- fan-assisted storage heaters which operate by radiant heat and convection;
- 'Electricaire' warm air units in which air is circulated through the heater by a fan and distributed directly, or via ducts.

7.13
The following paragraphs are extracted from the Electricity Association guidance:

> It should be recognized that those older storage heaters which use asbestos within their insulation are not entirely insulated with asbestos. Most of the thermal insulation in many of these appliances is either mineral wool or fibreglass which does not contain any asbestos. Various insulating boards containing asbestos such as 'Caposil' were used, typically where load bearing insulation was required. 'Caposil' has a density of between 200 and 600 kg/m^3 and some types contain 8–20% amosite or chrysotile. 'Caposil' is a trade name and encompasses a range of insulation material. Some varieties of 'Caposil', particularly those made after 1975, do not contain asbestos. Asbestos was also sometimes used for minor applications, such as the provision of small pads to insulate the feet, asbestos string and/or washers around electrical connections or graphitized asbestos as a dry packing in the fan motor, which are unlikely to present a health hazard. Asbestos was not used in the insulation of storage heaters manufactured after 1975 (except for Constor models).
>
> In determining the implications of asbestos insulation in these heaters, the following three situations were considered:

> i. Normal running
>
> Fibrous dust will be found in any domestic situation and tests using optical microscopy in dwellings with storage heaters have not shown any increase in fibre counts above the low levels expected in the domestic situation generally. The dust existing in the home is of course circulated more positively where fan controlled warm air heating units of any type are in use. Further tests were carried out using electron microscopy to identify the fibres. The results of tests did not show any statistically significant difference between asbestos fibre counts in homes with storage heaters and those without.

> ii. Repair of heaters
>
> The servicing or dismantling of heaters which use asbestos insulation materials should be carried out by properly trained and equipped staff. The majority of repairs to storage heating appliances can be carried out without disturbing insulation material containing asbestos. However, removal, repair or disturbance of asbestos insulation may only be undertaken in accordance with the Asbestos (Licensing) Regulations[5] the Control of Asbestos at Work Regulations 1987, their associated Approved Code of Practice (ACoP), and the ACoP that specifically relates to the removal of asbestos insulation, asbestos coating and asbestos insulating board.[2] Electricity Company staff are trained and equipped to work to the methods described in the above guidance and the Electricity Company or a suitably licensed contractor should be consulted before servicing and/or dismantling these units. Further advice on procedures may be obtained from the local area office of HSE.

> iii. Disposal of storage heaters
>
> At the end of their useful life storage heaters should be removed and disposed of taking appropriate precautions. Where practicable, they should be removed as complete units after sealing with suitable thick plastic sheeting, and if it is necessary to dismantle the heater, this should be undertaken by a contractor licensed under the Asbestos (Licensing) Regulations. Asbestos waste from storage heaters should be disposed of in accordance with the Special Waste Regulations 1980 made under the Control of Pollution Act (paragraphs 6.2 – 6.5).
> Individual householders who wish to dispose of storage heaters are advised to consult their local waste disposal department, and consult their local Electricity Company for general advice on the heaters and their repair.
>
> Heaters which are not included in the Electricity Association's list should be referred to the local Electricity Company who may be able to

Plate 1 Two beams, coated with sprayed amosite asbestos, in a ceiling void. The coating on the lower beam is in very poor condition.
The material has been damaged and is deteriorating. Debris has contaminated the glass fibre insulation on the left of the picture.

Plate 2 Sprayed amosite asbestos coating on a beam, in a ceiling void. The material is slightly damaged, showing evidence on indentation and scorching from work on an adjacent pipe connection. There are signs of deterioration around the angled bracket on right of the picture.

Plate 3 Chryostile asbestos pipe lagging in very poor condition. The material has substantially deteriorated.

Plate 4 Chrysotile asbestos boiler lagging cut away to allow a beam to be installed. The unsealed edges of the insulation are clearly exposed, although the bulk of the material is in good condition.

Plate 5 Damage to a small area of chrysotile asbestos pipe lagging. Some material is hanging loose and more material has been exposed. The bulk of the material is in good condition.

Plate 6 Slight damage to chrysotile asbestos lagging on a pipe elbow. Some debris has obviously fallen from the damaged area, but a small amount remains. The bulk of the material is in good condition.

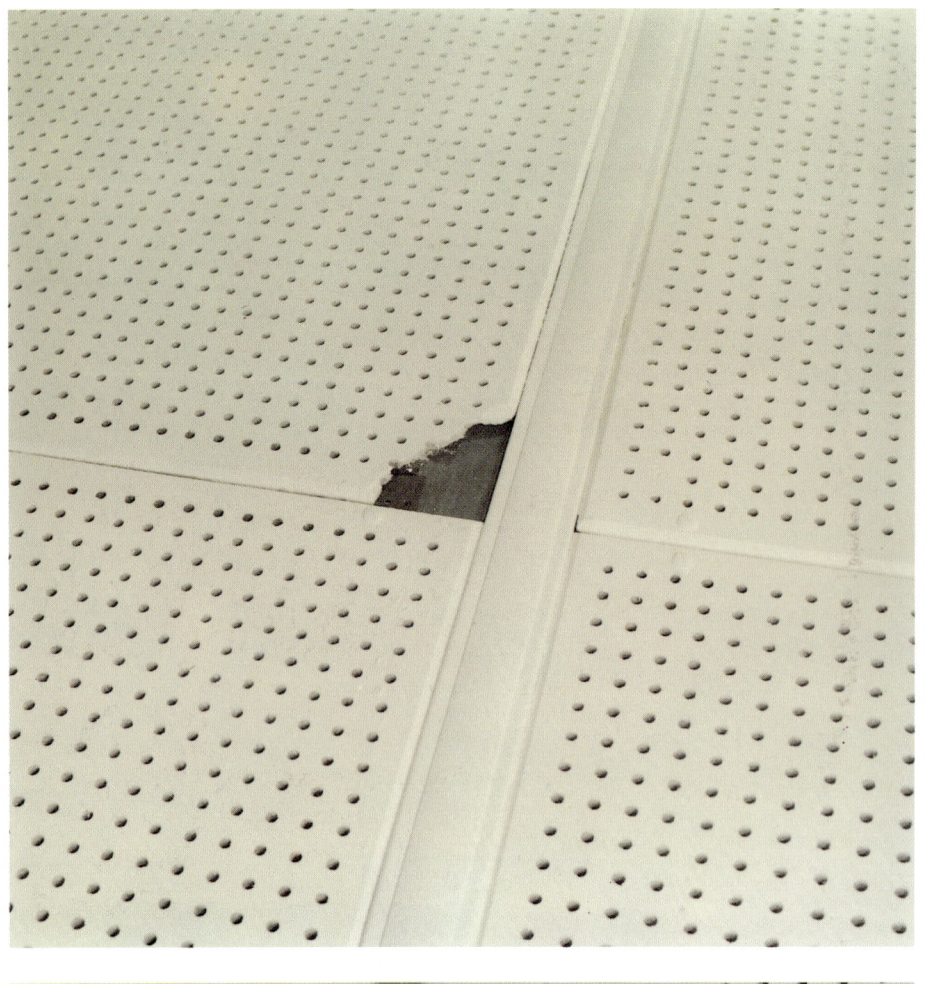

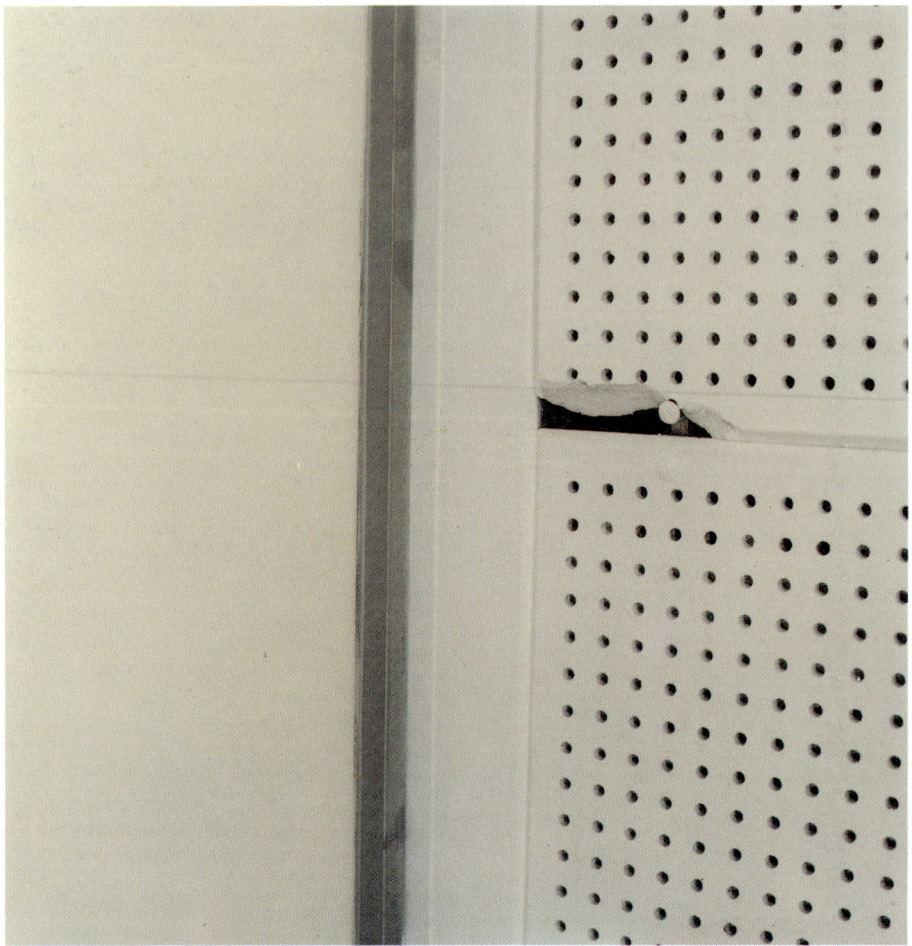

Plates 7 and 8 Corner damage to amosite asbestos insulating board ceiling tiles. While the boards are generally in good condition, such damage becomes increasingly common as old, brittle boards and tiles are disturbed, for example during maintenance work.

determine whether asbestos was used. If this cannot be established and the heaters are pre-1975, repair or dismantling should be dealt with on the assumption that they may contain asbestos. Householders are advised not to move or dismantle these heaters.

Do-it-Yourself – DIY work

7.14 Householders (among them many do-it-yourself enthusiasts) may be involved in handling a wide range of materials containing asbestos, either when they maintain the existing structure of their homes, garages, sheds or lean-to-roofs, or when they introduce new DIY products into the home for the same purpose. Some of the asbestos materials likely to be used are discussed in Section 3 of this booklet.

7.15 Before they plan their work programme DIY workers will need to be concerned with the safe handling of all materials containing asbestos. Above all, they should realise that asbestos fibres can only do harm if they are released into the air. Examples of work which can cause fibres to be released are, drilling or removing asbestos-cement sheets or structures, or, drilling or removing asbestos insulating board whilst carring out other work such as rewiring or plumbing. Anyone carrying out such work therefore, is advised to follow the practices and guidance for workers which has been prepared by HSE[15, 16]. Brief advice on asbestos-cement is given in paragraphs 3.9–3.16 of this booklet.

When working with asbestos materials there are certain key points to be followed:

do not drill, cut or disturb asbestos unless absolutely necessary;

take care not to create dust;

keep material wet;

clean up afterwards (not with a vacuum cleaner) and wash your protective clothing carefully;

keep the family away.

Before undertaking any work with asbestos materials the advice of the environmental health officer should always be sought.

7.16 The asbestos content of the materials in Annex 1 varies widely from 0.5–2% in putties to 90% or so for some wall plugging compounds. Once in place and during normal use, DIY products containing asbestos present very little hazard in the home; in most cases the fibres are bonded in the material and cannot be released unless the material is disturbed or damaged. However, during DIY work it is possible for some fibre release to occur, and precautions should be taken to avoid inhaling any dust. Asbestos materials should not be sanded down or scraped before painting and decorating. It may be better to paint or paper over existing decorations rather than attempting to remove them. Waste material and dust must be safely disposed of (as described in paragraphs 6.2 – 6.9).

7.17 Asbestos-free substitutes are now available for most products. Asbestos rope or string may still be stocked by specialist shops serving the plumbing and heating trade, and is used for caulking, packing and jointings. Although generally sold as dust-suppressed, this diminishes but does not eliminate the possibility of fibre release. In general the use of asbestos in DIY products is being phased out and few asbestos-containing materials are now available.

Labelling of asbestos products

7.18 The purpose of labelling is to enable consumers to identify products that contain asbestos so that they may take this into account before purchase. It also draws to the attention of persons having to work with these products the need to take adequate precautions. A European Community Directive[4] agreed in 1983 requires all asbestos products sold in member states to carry a standard label (illustrated in Annex 5). The label may be fixed to the asbestos product or to its packaging. The Directive also specifies a number of safety instructions to accompany products which are further processed or finished. The Directive is implemented in this country by the Asbestos Products (Safety) Regulations and the labelling requirements of the regulations came into effect on 20 March 1986. The European Directive and the regulations supersede a voluntary labelling scheme introduced by the UK asbestos industry in 1976.

8   Substitutes for asbestos

8.1   Asbestos-free substitutes are now available for many building materials which used to be made with asbestos. The characteristics of some of the substitute materials are described below, but this is not intended to be a comprehensive catalogue of alternatives to asbestos. Reviews of this subject have been written by Pye, and Hodgson. [25, 31, 32, 33] It should be noted that there is no single substitute for asbestos; the unique properties of asbestos depend upon its finely divided fibrous structure, which is what can make it hazardous to health. Substitutes for asbestos in the major areas of use in buildings including sprayed fire protection, thermal insulation and lagging, insulating board, fibre reinforced general building board, heat-resistant textiles, ropes and packings, sealants, mastics and roofing felts and floor coverings, are discussed in this section.

8.2   Many of the proposed substitutes are fibrous. These include natural organic fibres (cellulose and wool), synthetic organic fibres (polypropylene, polyvinylalcohol, aramid, polyimides, polyacrylonitrile), glass fibre, ceramic fibre and rockwool. Glass fibre in particular is widely used in thermal insulation and in fibre reinforced composites. Fibre dimensions of substitute fibres are generally larger than those of asbestos, but a proportion of rockwool and glass fibre materials are both respirable and inhalable.

8.3   When selecting substitutes for asbestos materials it is important to decide upon the performance needed for the particular use, for example resistance to fire, thermal conductivity and durability. Fire resistance is especially important – asbestos materials were often used because they were non-combustible, or had good fire resistance. When a specified period of fire resistance is required, the substitute material must be able to meet the specification.

Sprayed fire protection

8.4   Sprayed asbestos fire protection materials have been replaced by mineral fibre materials of several types. Mineral fibre sprayed materials are available for a range of maximum temperatures up to 650°C and for fire protection of up to 4 hours, depending upon thickness. High-temerature sprayed material made of ceramic fibres in an organic dry binder is available for temperatures up to 1200°C. Higher temperature needs are served by alumino-silicate fibres, alumino-silicate and chromia fibres, or alumina and zirconia fibres. A number of hard-set sprays based on vermiculite are available for thermal insulation and fire protection.

Thermal insulation and lagging

8.5   Ceramic fibres used in sprayed coatings are also available as bulk fibres, strips, assorted preformed shapes, felts and blankets. Their thermal stability at high temperatures is generally better than that of asbestos, but resistance to damage during handling or due to movement may not be as good. Rock wool, slag wool and glass fibre have been available for many years and have effectively replaced asbestos as thermal insulation in ordinary domestic and public buildings. Glass wools and mineral wools are normally used at temperatures ranging from subzero to about 500°C, although the material softens and melts about this temperature, and may lose strength around 300°C. Glass fibre may be treated with a vermiculite coating to maintain tensile strength and stability at higher temperatures.

8.6   Other insulating materials include polyurethane, polyisocyanurate and polyphenolic foams, calcium silicate, foamed glass and cellulose fibre. These materials are available as slabs or preformed shapes with a range of insulating properties. Preformed fibrous pipe lagging may be faced with aluminium foil, PVC or canvas. Many plastic foam insulation materials used for lagging are combustible and burn freely, producing smoke and fumes. Cellulose fibre also burns, unless it contains added flame retardants. Before a material is chosen, the fire performance required and consequences of using combustible material must be considered.

Insulating boards

8.7   Asbestos-free insulating boards were developed in the 1970s and asbestos-reinforced boards were phased out in the UK during 1980. A number of building boards reinforced with organic, inorganic and man-made fibres, with matrices of Portland cement, calcium silicates, vermiculite and perlite, together with filters and density modifiers, are available. For example, a range of cellulose fibre reinforced boards is available. These boards are intended as general cladding materials and as replacements for asbestos insulating board and 'shipboards'. Non-combustible boards containing vermiculite are available for fire protection of structural steel. Glass-reinforced gypsum boards made from glass fibre and a non-alkaline gypsum binder, are only suitable for interior use. Multilayered or composite boards incorporating insulants such as foamed glass, polyurethane, polystyrene beads or polyester resin with inorganic binder or particles, are also available. These boards usually contain added fire retardants.

Asbestos-cement

8.8 Substitute materials have been developed for the major uses of asbestos-cement – flat building boards, corrugated sheets, slates, moulded rainwater goods and low pressure pipes. Substitutes consisting of fibre reinforced cement are usually made by the traditional Hatschek process which will only accept short fibres and in the absence of asbestos, requires cellulose to be added so that a good felt is formed. The principal reinforcing fibres used are commonly polyvinyl alcohol (PVAL) or cellulose, which are well suited to the Hatschek process. At present most profiled sheeting and slates have a cement matrix reinforced with PVAL fibres but flat boards commonly consist of cellulose reinforced autoclaved calcium silicate and mica. Polyacrylonitrile (PAN) fibres are also sometimes used. To utilise longer fibres which can give rise to higher impact strengths, more complicated machinery is needed for example, the spray process used to produce flat alkali resistant glass fibre reinforced cement (GRC) board. More recent commercially made asbestos-cement replacement materials include those reinforced with polypropylene (PP), for example, PP fibres are now used to make a strong extruded slate. Fibrillated PP is used to reinforce corrugated sheeting and a new type of GRC corrugated sheet is currently also being produced.

8.9 Most substitutes for asbestos-cement described are safer to manufacture than the original product because the reinforcing fibres have large diameters and so are not respirable. Reinforcement is less efficient than with asbestos, partly because of the large fibre diameter, so the new products are mostly less strong. Fibres are probably less durable than asbestos and organic fibres are less temperature resistant. Long term durability is difficult to assess for new materials. Fibre specific accelerated tests have only so far been developed for glass fibre and PP reinforced cement products. Many failures occurred with some of the substitutes tried in the early 1980s. On the other hand a trial of the commonest types of non-asbestos fibre reinforced cement material has shown little loss of strength during 5 years natural weathering. Products made in the last 3 years appear in many cases to be performing satisfactorily. Many other substitutes are available for asbestos-cement moulded rainwater goods, including non-fibrous PVC (polyvinylchloride), or metal products; the latter may also be used for flue pipes.

8.10 Alternatives to fibre-reinforced cement sheets are sheet metals and glass-reinforced plastics. Aluminium and steel sheets in a variety of finishes or colour coatings, are available as roof and wall claddings, tiles and slates and moulded goods. Coated steel profiles may have a relatively short guarantee for the protective coating, and consideration should be given as to whether the coating material could cause environmental problems when recycling the steel after use. Glass-reinforced plastic (GRP) sheets may be used where resistance to combustion is not needed, but GRP has poor scratch resistance. The fire propogation properties of GRP can vary widely depending upon composition and surface finish and this must be taken into account when they are used. In general, thermosetting plastics char on heating and burn, some producing large amount of smoke and noxious fumes. Thermoplastics melt, releasing molten droplets and smoke.

Other products

8.11 Flame or heat-resistant textiles are made using glass, ceramic and various polymer filaments and are available for a wide range of temperatures. Glass fibre fabrics may be used as fire curtains, although the upper limit on working temperature of 550°C means that the material may not be suitable for protection from fire or molten metal splashes. Glass fibre may be reinforced to improve resistance to abrasion. Glass fibre rope available as a replacement for asbestos rope, and chopped strand glass fibre may be formed into non-woven papers and felts for packings. Higher temperature packings are available in ceramic or carbon fibre. For lower temperature applications, flame-resistant wool (normally aluminized for fire resistance) leather or nylon, may be suitable as replacement for asbestos textiles in protective clothing. A British Standard (BS 6575) for light and heavy duty fire smothering blankets which do not contain asbestos, was issued in 1985.

8.12 Asbestos-free sealants and mastics are now available reinforced with other fibres. PTFE (polytetrafluorethylene) can be used as a low temperature jointing and packing material. Asbestos-free textured plasters are also available and are claimed to give comparable covering power to the asbestos versions.

Health aspects of mineral fibre substitutes

8.13 Mineral fibre substitutes are generally believed to be safer than asbestos, but mineral fibres implanted in experimental animals have been shown to cause mesothelioma.[34] A World Health Organization meeting in 1982[35, 36] on the biological effects of man-made mineral fibres (MMMF) noted that there was no clear evidence of increased mortality from cancer or other diseases associated with the occupational exposure to MMMF in manufacturing plant. However, occupational exposure to MMMF in manufacturing plant were considerably lower, sometimes by one or more orders of magnitude, than the exposure of asbestos workers to asbestos. Furthermore, very few workers have been followed for as long as 20 to 30 years from first employment, which is the usual latency period for lung cancer or mesothelioma after exposure to asbestos. Although there is some evidence that the number of deaths from lung cancer is higher than expected, the numbers involved are very small, and there may also have been exposure to asbestos. Occupational studies are being continued. On the available evidence, the effects of exposure to MMMF cannot be determined with certainty, but there is evidence that the biological effects of different types of MMMF (and other natural mineral fibres) vary widely, and that both fibre diameter and the physico-chemical nature of the material are relevant factors.

8.14 In 1985 the Department of Health's Committee on Carcinogenicity was asked to review all the available evidence for carcinogenic hazard from workplace exposure to MMMF. The Committee's advice given in December 1987 may be summarised as:

(1) Current evidence indicates that exposure to MMMF increases the lung cancer risk in humans.

(2) The risk is probably only for bronchial carcinoma in man as no excess of mesothelioma has been observed in exposed workers.

(3) Experimental data suggests that the magnitude of the risk might be related to fibre size and shape but the exact relationship to fibre dimensions has not been determined.

(4) Despite the uncertainty over the precise dose-response relationship between exposure and lung cancer, the Committee though it would be prudent to treat inhalation of MMMF in the same way as chrysotile asbestos.

The Committee on Carcinogenicity's recommendation to HSE was that:

'It would be prudent to act on the basis that sufficient exposure to any form of man-made mineral fibre in the production or user industries may increase the risk of lung cancer amongst the workforce. The information available is insufficient to indicate whether there is any particular level at which no adverse effect would be apparent.'

8.15 The Committee also advised that fibre-count exposure limits should be considered for all types of MMMF, and its recommendations are the basis of UK legislative controls set out in Schedule 1 of the Control of Substances Hazardous to Health Regulations 1988. These apply to all forms of MMMF, and Schedule 1 of the Regulations assigns to them a Maximum Exposure Limit (MEL). The MEL for MMMF is expressed in two ways: gravimetrically as a limit of 5 milligrams per cubic metre of air, and as an airborne fibre limit of two fibres per millilitre. Both measurements are averages obtained from samples taken over an 8-hour period.

8.16 A working party (associated with the Advisory Committee on Asbestos) on the health effects of MMMF, made the following recommendation on non-occupational exposure:[37]

'Non-occupational exposures to man-made mineral fibre are likely to be of short duration and carry no significant risk to health'. They also recommended that 'Care should be taken in the design of plant and insulation in buildings to prevent the distribution of fibres through ducted air distribution systems.'

The HSE has published guidance on the possible health risks which could result from exposure to MMMF and the precautions which may be needed.[38]

8.17 In 1987 the Department of Health's Committee on Carcinogenicity was asked by DoE to assess the risk to health from domestic MMMF loft insulation, DIY installation and subsequent disturbance in the home.

The Committee advised that:

the levels of exposure to MMMF reported for living spaces resulting from domestic loft insulation do not pose a carcinogenic risk of any practical consequence to the health of residents;

the infrequent and short term exposure to the higher levels of MMMF associated with DIY installation or disturbance of insulation do not pose a significant additional risk;

it would be prudent nonetheless for installers to wear an appropriate mask as recommended during installation;

the overall situation should be kept under review in the light of further developments in materials and types of installation.

In view of the Committee's advice members of the public working with MMMF loft insulation are advised:

to wear a mask conforming to British Standard BS 6016 or European Standard PrEN 149. Mask to BS 2091 or European Standard EN140 with an appropriate filter, are also suitable, although more expensive. **A mask normally used for 'nuisance dust' is not suitable;**

that mineral wool is a skin as well as a bronchial irritant, so rubber gloves and smooth clothing or overalls should be worn to avoid picking up loose fibres. Clothing or overalls should be tucked into gloves and socks;

that after use the mask should be put in the dustbin. Overalls, gloves and other clothing should be washed before future use;

and that if working with MMMF in a loft:

to ensure that the loft is well lit;

to avoid spreading insulation around the house, by putting waste insulation or offcuts into bags while in the loft, before disposal.

This advice was made publicly available and circulated to local authorities in May 1988.

Annex 1

Asbestos materials used in buildings

Asbestos product	Use	Asbestos content	Remarks
Sprayed asbestos coatings.	Thermal and acoustic insulation. Fire and condensation protection.	Sprayed coatings contain up to 85% asbestos. A mixture of types was used until 1974. Crocidolite was used for the thermal insulation of steam turbines until 1970. Amosite was used for fire protection of structural steel condensation protection and acoustic control. Chrysotile, mixed with mineral wool and binder, was used until 1974. Chrysotile was also used as a coating on top of other sprayed asbestos.	Potential for fibre release unless sealed. Potential increases as the materials age or become friable and disintegrate. Dust released may then accumulate. Removal of sprayed coating is a licensed activity.
Asbestos lagging.	Thermal insulation of pipes, boilers, pressure vessels, preformed pipe sections, slabs, tape, rope, corrugated paper, quilts, felts and blankets.	All types of asbestos have been used. Content varies (eg. 6–8% in calcium silicate slabs, 100% in blankets, felts etc.).	Friability depends on the nature of the lagging. Potential for fibre release unless sealed. Potential increases as the materials age or become friable and disintegrate. Dust released may then accumulate. Removal of lagging is a licensed activity.
Insulating boards.	Fire protection, thermal and acoustic insulation, resistance to moisture movement and general building board. Used in ducts, firebreaks, infill panels, partitions and ceilings (including ceiling tiles), roof underlays, wall lining, bath panels, external canopies and porch linings.	Crocidolite used for some boards up to 1965. 16–40% amosite or a mixture of amosite and chrysotile.	Likely to cause a dust hazard if very friable, broken, abraded, sawn or drilled.
Insulating board cores and linings of composite products.	Acoustic attenuators cladding infill panels, domestic boiler casings, partition and ceiling panels, oven linings and suspended floor systems.		

Asbestos materials used in buildings *(cont.)*

Asbestos product	Use	Asbestos content	Remarks
Ropes and yarns.	Lagging. Jointing and packing materials. Heat/fire resisting gaskets and seals. Caulking in brickwork. Boilier and flue sealing. Plaited asbestos tubing in electric cable.	All types of asbestos where used until about 1970. Since then only chrysotile has been used. Asbestos content approximately 100%.	Fibre may be released when large quantities of unbonded material are stored or handled. Caulking etc. in situ is not likely to release fibre.
Cloth.	Thermal insulation and lagging including fire-resisting blankets, mattresses and protective curtains, gloves, aprons, overalls etc. Curtains, gloves, etc., were sometimes aluminized to reflect heat.	All types of asbestos have been used in the past. Since the mid 1960s, the vast majority has been chrysotile. Asbestos content approaching 100%.	Fibres may be released if material is abraded.
Millboard and paper.	General heat insulation and fire protection. Electrical/heat insulation of electrical equipment and plant. Asbestos-paper has been used in the manufacture of roofing felt and damp-proof courses, steel composite wall cladding and roofing, (see asbestos-bitumen products below), vinyl flooring, facing to combustible boards, flame resistant laminate and corrugated pipe insulation. Millboard was used in laboratories for thermal insulation.	Crocidolite was used in some millboard manufactured between 1896–1965; subsequently chrysotile. Asbestos content approximately 100%.	Uncoated asbestos paper and millboard is not highly bonded and should not be used where subject to abrasion or wear.

Asbestos materials used in buildings *(cont.)*

Asbestos product	Use	Asbestos content	Remarks
Asbestos-cement.			
• Profiled sheets.	Roofing. Wall cladding and weather-boarding.	10–15% asbestos (some flexible boards contain a small proportion of cellulose). Crocidolite and amosite have been used in the manufacture of asbestos-cement products, although chrysotile is the most common type.	Likely to release fibres if abraded, handsawn or worked on with power tools, cleaned with high power hoses, deteriorated or decomposed.
• Semi-compressed flat sheet and partition board.	Partitioning in farm buildings and housing, shuttering in industrial buildings, decorative panels for facings, bath panels, soffits, linings to walls and ceilings, portable buildings, propagation beds in horticulture, domestic structural uses, fire surrounds and composite panels for fire protection.		
• Fully compressed flat sheet and partition board.	As above but where stronger materials are required.		
• Tiles and slates (made from fully compressed flat sheet).	Cladding. Decking and promenade tiles. Roofing.		
• Preformed moulded products.	Cisterns and tanks, drains, sewer pipes and rainwater goods. Flue pipes. Fencing. Roofing components (fascias, soffits etc.). Cable troughs and conduits. Ventilators and ducts. Window/flower boxes.		
Asbestos-bitumen products.	Bitumen roofing felt. Bitument damp-proof course (dpc). Semi-rigid asbestos-bitumen roofing. Gutter linings and flashings. Asbestos-bitumen coatings on metals.	Chrysotile fibre or asbestos paper (approximately 100% asbestos) in bitumen.	Fibre release unlikely during normal use. Roofing felts, dpc and bitumen based sealants must not be burnt after removal.

Asbestos materials used in buildings *(cont.)*

Asbestos product	Use	Asbestos content	Remarks
Flooring.	Thermoplastic floor tiles.	Up to 25% asbestos.	Fibre release is unlikely to be a hazard under normal service conditions. Fibre may be released when material is cut, and there may be substantial release when flooring, particularly with paper backing, is removed.
	PVC vinyl floor tiles and unbacked PVC flooring.	Normally less than 10% chrysotile.	
	Asbestos paper backed PVC floors.	Paper backing approximately 100% chrysotile asbestos.	
Textured coating.	Coatings on walls and ceilings.	3–5% chrysotile asbestos.	Fibres may be released when 'dry mix' materials are prepared or when old coating is rubbed down. The material must **not** be power sanded. Remove by wet scraping if necessary.
Mastics, sealants, putties and adhesives.	General.	0.5%–2%.	The only possible hazard is from sanding of hardened material. Sanding down with power tools should be avoided.
Reinforced PVC and plastics.	Panels and cladding. Reinforcement for domestic goods.	Variable.	
Wall plugging compound.	Wall fixings.	Greater than 90%.	Made up from loose asbestos and cotton fibre with plaster dust.

Asbestos materials used in buildings *(cont.)*

Asbestos in domestic appliances	Asbestos material	Asbestos content	Remarks
Hairdriers, fan and radiant electric heaters, irons, toasters, washing machines, tumble driers, spin driers, dish washers, refrigerators and freezers.	Paper, element formers, brake pads, compressed fibre gaskets and seals, rubberised or other polymer gaskets and seals.	Variable.	Asbestos paper has been used for heat insulation in hair driers. In general, gaskets and brake pads are sealed within appliances and are unlikely to release fibre into the atmosphere.
Cookers.	Insulating board. Fire cement, compressed fibre seals, rubberised or other polymer seals.	16–40%. Variable.	
Simmering mats.	Millboard.	Approaching 100%.	
Iron stands.	Paper, millboard and asbestos-cement.	Approaching 100% and 10–15%.	
Catalytic gas heaters.	Compressed asbestos fibre panels.	100%, sometimes covered by a glass fibre mesh.	Those in doubt about the condition of their heater should contact their local authority environmental health department.
Gas warm-air heaters.	Aluminium backed paper, cloth and insulating board.	Approaching 100% and 16–40%.	
Boilers, pipework.	Asbestos/plaster with or without a surface fibre layer.	Variable.	
Electric 'warm-air' and storage heaters.	'Caposil' insulating blocks. Insulating board. Paper, string, compressed fibre washers, rubberised/polymer bonded washers.	Variable. 16–40%. Variable.	
Radiators.	String, washers.	Approaching 100%, variable.	

Annex 2
Sampling asbestos materials

Spray coating. Different mixtures containing different materials may have been used in different areas and layers. Material may also have been removed, repaired or patched at various times. Samples should be taken by carefully removing pieces of approximately 5 cm^2, at a rate of approximately one sample per 10–15 m^2 or, in installations exceeding 100 m^2, one sample per 25–30 m^2. At least one sample should be taken from each patched area. Care should be taken to include all layers of sprayed coating through to the covered surface.

Asbestos lagging. Different mixtures containing different materials may have been used in different areas and layers. Material may also have been removed, repaired or patched at various times. For example, asbestos materials may have been stripped from long 'runs' of pipe, but have been left around pipe elbows, taps and valves. In general, one sample should be taken per three metre run of pipe, with particular attention paid to different layers and the areas specifically mentioned. For long pipe runs (over 20 metres) one sample per six metres will usually be sufficient. Samples should be taken as a 'core' of approximately 5 cm^2 cross section, to include all material between the inner and outer surfaces. At least two samples of boiler lagging should be taken on any one unit. Additional samples should be taken from each separate 'patched' area.

Insulating board. Only one sample (of approximately 5 cm^2) per sheet should be necessary, provided that the sample is representative of the sheet as a whole. If possible, samples should be taken near the edge where the board is likely to be backed by a batten (this makes repair with filler much easier). The material may have been used as a core or lining to other products, such as fire doors or metal ducting (Annex 1). If numerous seemingly identical pieces or sheets have been used in one area, two or three sheets should be sampled; if they contain asbestos, the others can be assumed to do so.

Asbestos ropes and yarns. May be covered by fillers, cement, etc., in use and the process of obtaining the sample may render the material useless as a sealant or as caulking. A single small piece of rope or yarn will be sufficient for identification.

Asbestos cloth. A small fragment should be sufficient for analysis.

Millboard and paper. This is a uniform material, so where isolated sheets or suspect material are available, one sample per sheet will be sufficient. Millboard may also have been used as a 'patching' material in other circumstances.

Asbestos-cement products. Asbestos-cement is a uniform material and unless there are obvious differences between sheets, pipe runs, etc., suggesting that different materials have been used at different times, two or three samples (approximately 5 cm^2) should be sufficient for a roof or a 'run' of guttering or pipe work. It should be noted that slates are often coloured or coated. Particular care should be taken to avoid accidents when samples are taken from roofing material. Crawl boards must be used if it is necessary to take samples on a roof.

Bitumen roofing felt, damp-proof course, gutter lining and flashings. One small sample per roll or run of felt or other material should be sufficent.

Textured coatings. These were often mixed on site and may be non uniform. Two to three samples should be taken in different areas of the ceiling or coated area.

Thermoplastic floor tiles. One sample from one tile of each colour used in each room or location where they are laid.

Unbacked PVC flooring. One sample per roll, or per room.

'Backed' cushion floor or embossed floor covering. These materials may be backed with asbestos paper, so samples should only be taken from the backing material. One sample per roll, or per room, should be sufficient.

Annex 3

Asbestos assessment charts

Refer to the Control of Asbestos at Work Regulations 1987 and the Approved Code of Practice[6] before any work on materials which *may* contain asbestos, is carried out.

Chart 1: Asbestos materials

Refer to Sections 4 and 5.

Notes

1. The Control of Asbestos at Work Regulations 1987 (Reg. 4) require that an employer shall not carry out any work which exposes or is liable to expose any of his employees to asbestos unless he has made an adequate assessment of that exposure. Furthermore the employer is obliged to identify the type of asbestos involved in the work, or assume that it is crocidolite or amosite and for the purposes of the Regulations, treat it accordingly.

2. The purpose of an assessment (Reg. 5), which should generally be in writing, is to enable a correct decision to be made about the measures necessary to control exposure to asbestos. If the assessment concludes that exposure is liable to exceed the action level (Reg. 2) then other provisions of the Regulations will apply. The assessment also enables the employer to satisfy himself and to demonstrate to others that all the factors pertinent to the work have been considered, and that an informed and correct judgement has been reached about the risks and the steps which need to be taken to achieve and maintain adequate control.

3. Where material is in good condition but is or will become highly vulnerable to damage, management alone may not be sufficient to prevent a hazard. Treat the material as not in good condition.

4. Insulating board was frequently used as a general building board and visually may be confused with plasterboard or flat asbestos-cement sheet. Bulk samples will distinguish insulating board from plasterboard. Asbestos-cement was normally made with chrysotile asbestos and insulating board with amosite asbestos, but all types of asbestos have been used, in varying proportions, in both products. Insulating board was frequently nailed in position, asbestos-cement is often fixed with screws or bolts.

Some types of 'Caposil' insulation blocks, found in some storage and warm air heaters manufactured prior to 1976, contain asbestos. For the purposes of the assessments charts, these blocks should be treated like insulating board. However, work with those blocks which have a density of less than 500 kg/m^3, will be subject to the Asbestos (Licensing) Regulations.[5]

The risk of fibre release from insulating boards is very low, except when they are worked, or when the material is damaged. It is important that a note should be made on plans and records and a label attached where maintenance work is liable to cause frequent contact with the material.

5. If it is necessary to disturb asbestos frequently, the cost of precautions required may make it more cost effective to remove the material. In housing, occupants, especially those in public sector rented accommodation, should be made aware of the location of any asbestos materials and advised of appropriate precautions.

Chart 1: Asbestos materials

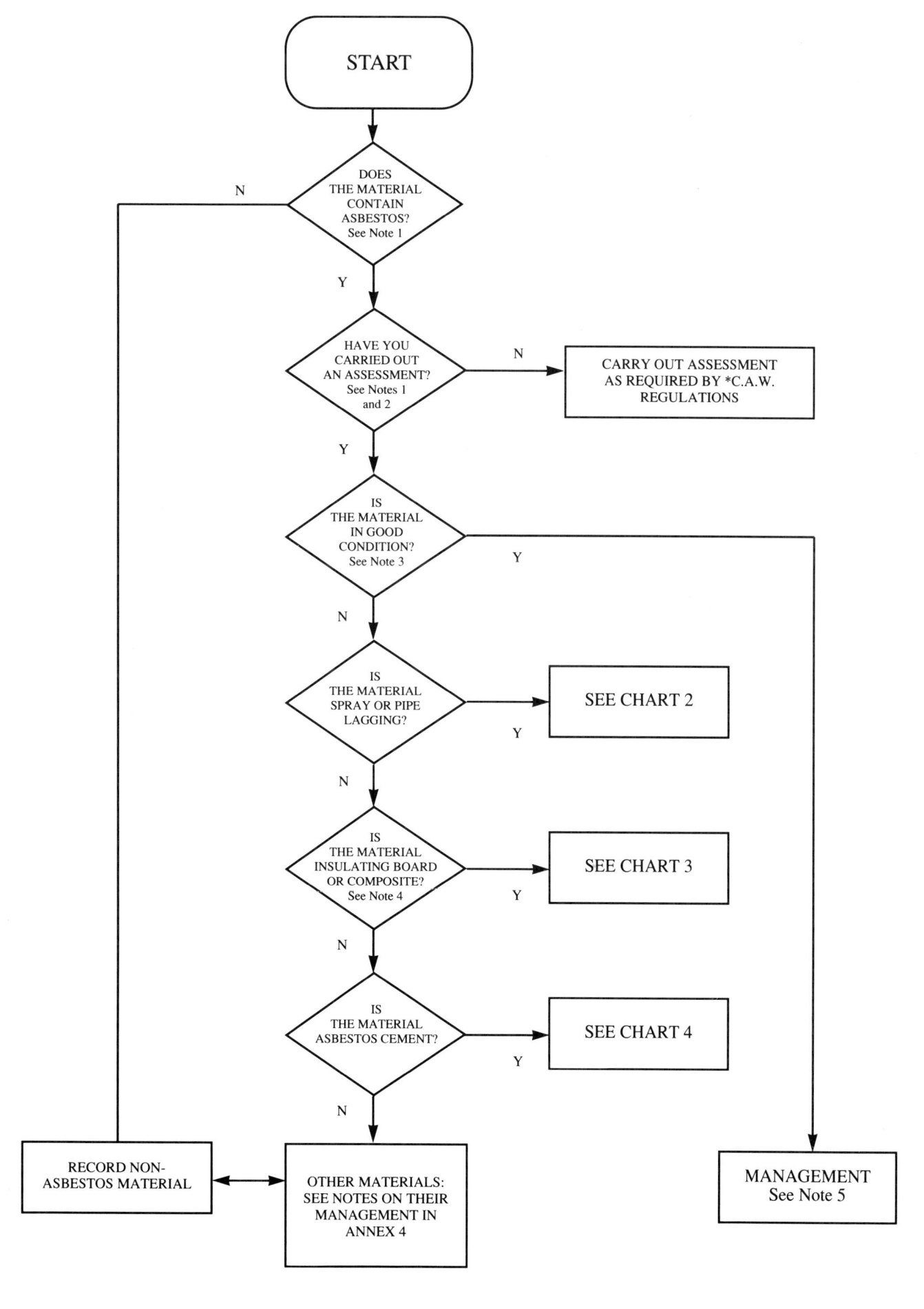

```
                            START
                              │
                              ▼
              ┌───────────────────────────────┐
      N       │ DOES THE MATERIAL CONTAIN     │
    ◄─────────│ ASBESTOS?  See Note 1         │
    │         └───────────────────────────────┘
    │                         │ Y
    │                         ▼
    │         ┌───────────────────────────────┐     N     ┌──────────────────────────┐
    │         │ HAVE YOU CARRIED OUT AN       │──────────►│ CARRY OUT ASSESSMENT AS  │
    │         │ ASSESSMENT? See Notes 1 and 2 │           │ REQUIRED BY *C.A.W.      │
    │         └───────────────────────────────┘           │ REGULATIONS              │
    │                         │ Y                         └──────────────────────────┘
    │                         ▼
    │         ┌───────────────────────────────┐     Y
    │         │ IS THE MATERIAL IN GOOD       │─────────────────────────────────────┐
    │         │ CONDITION? See Note 3         │                                     │
    │         └───────────────────────────────┘                                     │
    │                         │ N                                                   │
    │                         ▼                                                     │
    │         ┌───────────────────────────────┐     Y     ┌──────────────────────────┐
    │         │ IS THE MATERIAL SPRAY OR      │──────────►│ SEE CHART 2              │
    │         │ PIPE LAGGING?                 │           └──────────────────────────┘
    │         └───────────────────────────────┘
    │                         │ N
    │                         ▼
    │         ┌───────────────────────────────┐     Y     ┌──────────────────────────┐
    │         │ IS THE MATERIAL INSULATING    │──────────►│ SEE CHART 3              │
    │         │ BOARD OR COMPOSITE? See Note 4│           └──────────────────────────┘
    │         └───────────────────────────────┘
    │                         │ N
    │                         ▼
    │         ┌───────────────────────────────┐     Y     ┌──────────────────────────┐
    │         │ IS THE MATERIAL ASBESTOS      │──────────►│ SEE CHART 4              │
    │         │ CEMENT?                       │           └──────────────────────────┘
    │         └───────────────────────────────┘
    │                         │ N
    ▼                         ▼                                                     ▼
┌─────────────────┐  ┌───────────────────────────┐                       ┌──────────────────┐
│ RECORD NON-     │◄─►│ OTHER MATERIALS: SEE      │                       │ MANAGEMENT       │
│ ASBESTOS        │   │ NOTES ON THEIR MANAGEMENT │                       │ See Note 5       │
│ MATERIAL        │   │ IN ANNEX 4                │                       │                  │
└─────────────────┘   └───────────────────────────┘                       └──────────────────┘
```

*Control of Asbestos at Work Regulations 1987

Chart 2: Sprayed asbestos and lagging

Refer to Sections 4 and 5.

Notes

1. The chart deals with material which is considered not to be in good condition. All work on these products falls within the scope of the Asbestos (Licensing) Regulations 1983,[5] and must be carried out in accordance with the HSE Code of Practice on 'Work with asbestos insulation, asbestos coating and asbestos insulating board'[2], and generally undertaken by a licensed contractor.

2. To be readily repairable, damage to the installed material must be slight.

Repair work should be restricted to:
- trowelled repairs or patching of small areas of the asbestos material;
- applying small areas of sealant;
- making good slight damage to boxing.

Repairs should be carried out taking the appropriate precautions, and observing the Control of Asbestos at Work Regulations 1987, Code of Practice and HSE guidance.

3. The term accessible is explained in paragraph 4.11.

4. Accessible material which is not extensively damaged will probably need protection against further damage, and sealing or enclosure may be necessary.

5. Loose debris and quantities of material detached from the main body of the asbestos may indicate that the asbestos is breaking up and highly friable. If there is no evidence of loose debris and the asbestos is firmly bonded to the substrate, it may be sealed (paragraph 5.5), or enclosed (paragraph 5.6). Sprayed coatings and lagging can be sealed with sprayed or bituminous coating, or with a hard setting cement-type coating. If necessary, cement coatings may be supported by metal mesh. Sealed sprayed coatings may be vulnerable to water damage, particularly where these are on the underside of flat roofs.

6. Enclosure may not be feasible if the area involved is very large, in long roofing structures for example, or where access to the asbestos material is restricted. If the enclosure would be vulnerable to damage, if access is needed for maintenance and repair, or enclosure is not feasible, then the asbestos must be removed.

7. When sprayed coatings or laggings are removed (paragraph 5.7), it will be necessary to empty the building or seal off the working area. The whole area should be thoroughly cleaned up afterwards. As it is not usually possible to remove all traces of asbestos, a sealing coat should be applied after removal. After removal work, the airborne fibre concentration should be measured before the area is reoccupied, using procedures specified in HSE Guidance Note EH 10[18]

8. In housing, occupants, especially those in public sector rented accommodation, should be made aware of the location of any asbestos materials and advised of appropriate precautions.

Chart 2: Sprayed asbestos and lagging

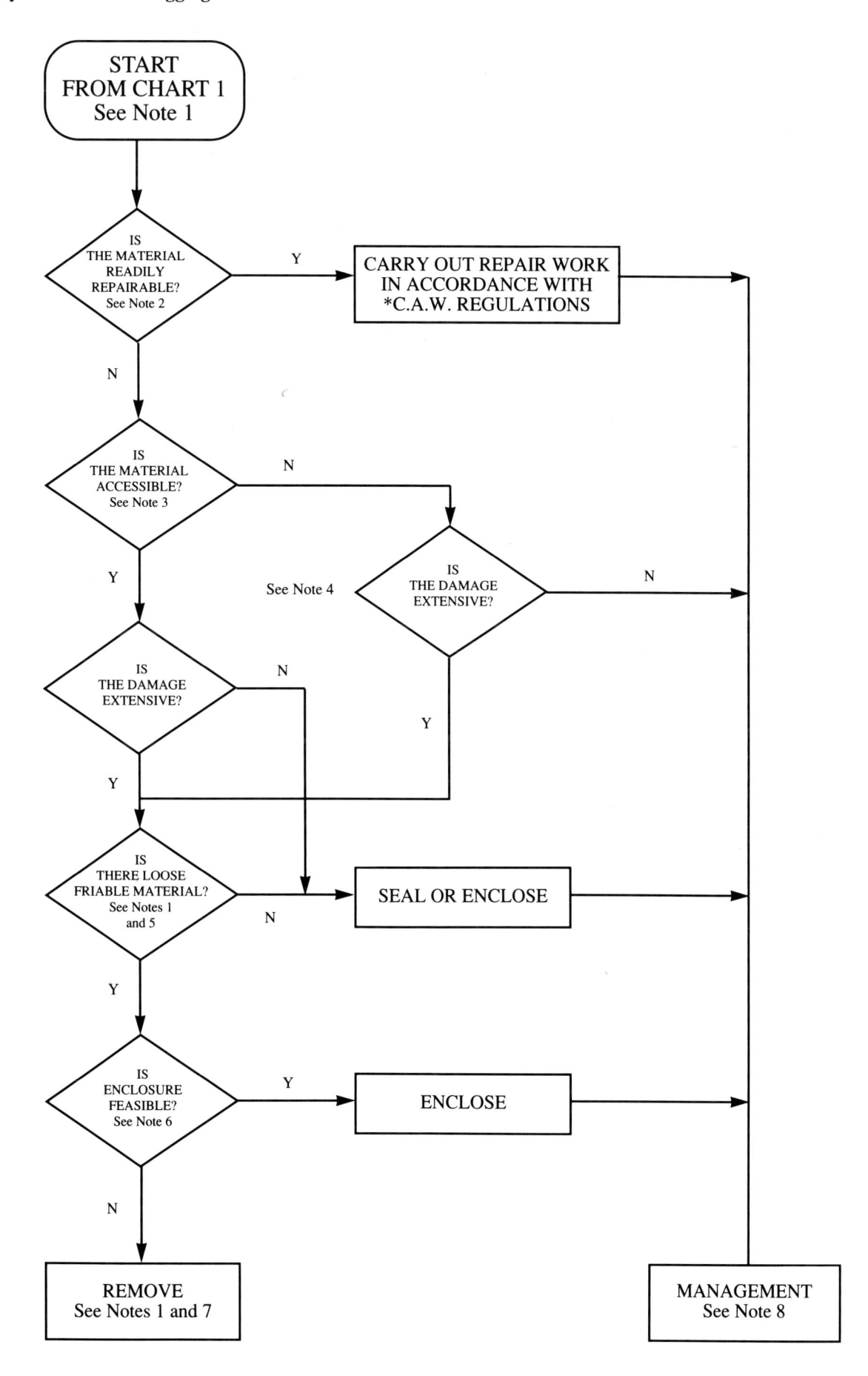

```
                    ┌─────────────────┐
                    │     START       │
                    │ FROM CHART 1    │
                    │   See Note 1    │
                    └────────┬────────┘
                             │
                             ▼
                    ╱ IS THE MATERIAL ╲      Y      ┌──────────────────────┐
                    ╲   READILY        ╱──────────▶│ CARRY OUT REPAIR WORK│
                    ╱   REPAIRABLE?    ╲           │ IN ACCORDANCE WITH   │
                    ╲   See Note 2    ╱            │  *C.A.W. REGULATIONS │
                             │                     └──────────────────────┘
                             │ N
                             ▼
                    ╱  IS THE MATERIAL ╲     N
                    ╲   ACCESSIBLE?     ╱─────────┐
                    ╱   See Note 3     ╲          │
                             │                    ▼
                             │ Y          ╱ IS THE DAMAGE ╲   N
                             │            ╲   EXTENSIVE?   ╱─────────▶
                             │            See Note 4
                             ▼                    │ Y
                    ╱  IS THE DAMAGE  ╲    N      │
                    ╲   EXTENSIVE?     ╱──────┐   │
                             │                │   │
                             │ Y              ▼   ▼
                             │         ┌──────────────────┐
                             │         │ SEAL OR ENCLOSE  │
                             ▼         └──────────────────┘
                    ╱ IS THERE LOOSE  ╲
                    ╲ FRIABLE MATERIAL?╱
                    ╱ See Notes 1 and 5╲
                             │
                             │ Y
                             ▼
                    ╱   IS ENCLOSURE   ╲    Y     ┌──────────┐
                    ╲    FEASIBLE?      ╱───────▶│ ENCLOSE  │
                    ╱   See Note 6     ╲         └──────────┘
                             │
                             │ N
                             ▼
                    ┌──────────────────┐          ┌──────────────────┐
                    │     REMOVE       │          │   MANAGEMENT     │
                    │ See Notes 1 and 7│          │   See Note 8     │
                    └──────────────────┘          └──────────────────┘
```

*Control of Asbestos at Work Regulations 1987

Chart 3: Asbestos insulating board, insulating blocks and composite products

Refer to Sections 4 and 5.

Notes

1. The chart deals with material which is considerd not to be in good condition. All work with asbestos is controlled by the Control of Asbestos at Work Regulations 1987 but certain types of product, such as asbestos coating and asbestos insulation, also fall within the scope of the Asbestos (Licensing) Regulations 1983[5]. Most asbestos insulating board has a density of approximately 700 kg/m^3. If crocidolite asbestos is present, the HSE must be notified 28 days before work begins. Any work on insulating board should follow HSE Guidance Note EH 37[15].

2. To be readily repairable, damage to board must be slight. Surface scratches may be sealed or painted, breaks taped and small punctures patched with filler. If the board is not covered it may be painted or otherwise sealed as a precaution against light abrasion.

3. The term accessible is explained in paragraph 4.11.

4. The material may be sealed (paragraph 5.5) by painting with an initial coat of diluted PVA emulsion followed by one or more full strength coats. The surface should be prepared and damaged material repaired where possible (see Note 2), but the material should not be sanded or wire brushed. Dusty surfaces can be cleaned with a suitable industrial vacuum cleaner, fitted with a high efficiency filter,[15] or wiped with a damp cloth (which should be sealed in a plastic bag whilst still damp). A domestic vacuum cleaner must not be used. Sealing does not protect the material from more violent impact. Covering the board with hardboard, plasterboard or a similar material may be preferred.

5. If the material is very badly damaged; is very extensive in area, or is subject to frequent violent impact, then sealing or enclosure (paragraph 5.6) may not be feasible.

6. Removal (paragraph 5.7) of large areas of asbestos insulating board should be carried out by trained staff or a specialist contractor. Although it may not be necessary to empty a building, the working area should be segregated and people not engaged in the work should be kept out of it. When asbestos insulating board is removed, it should be wetted to suppress dust and sheets should be removed whole, not broken up. Replacement board must have equivalent fire performance where this is required.

7. In housing, occupants, especially those in public sector rented accommodation, should be made aware of the location of any asbestos materials and advised of appropriate precautions.

Chart 3: Asbestos insulating board, insulating blocks and composite products

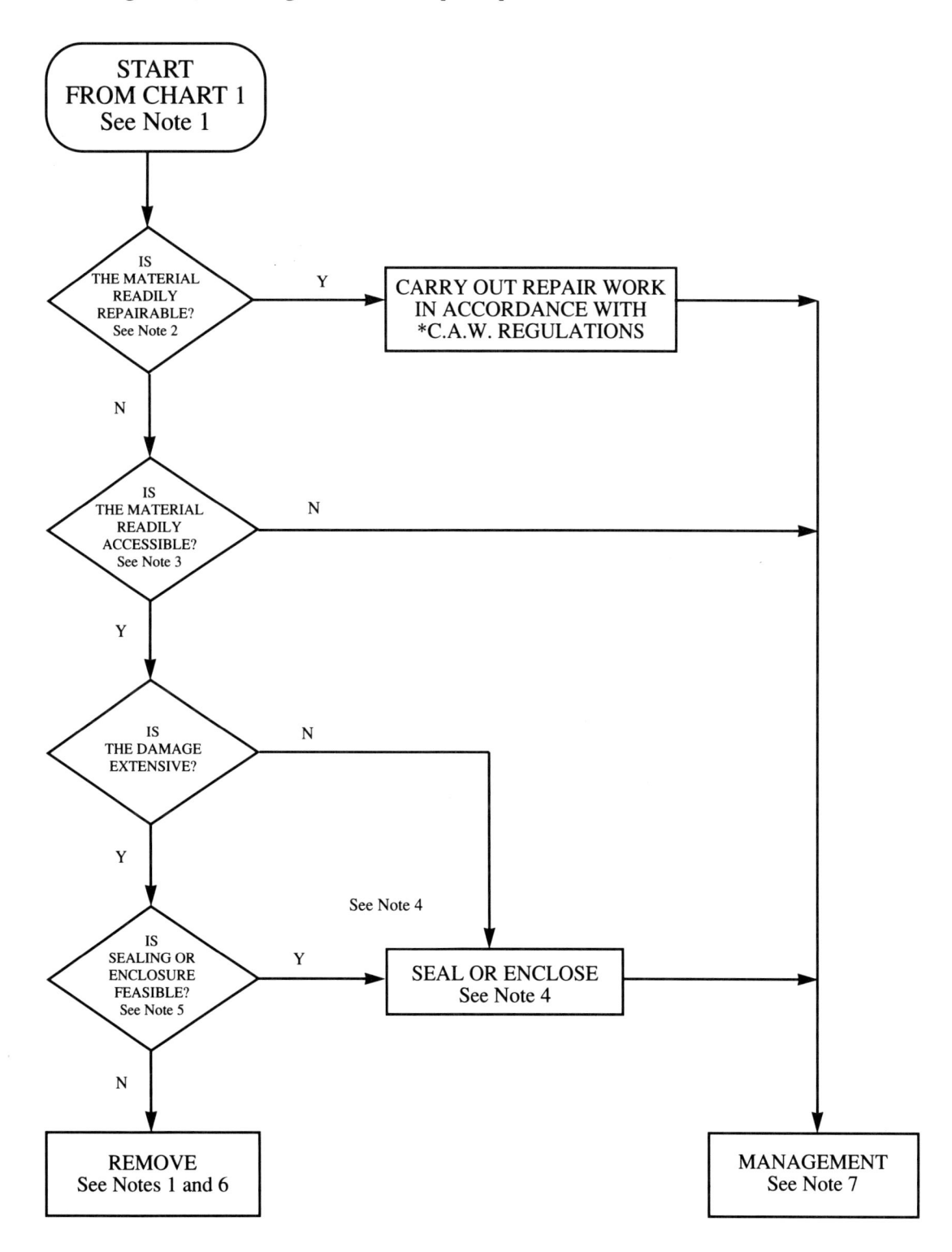

```
              START
         FROM CHART 1
           See Note 1
                │
                ▼
        ┌───────────────┐
        │      IS       │         Y      ┌──────────────────────┐
        │ THE MATERIAL  │──────────────▶│ CARRY OUT REPAIR WORK │
        │   READILY     │                │   IN ACCORDANCE WITH │
        │  REPAIRABLE?  │                │   *C.A.W. REGULATIONS│
        │   See Note 2  │                └──────────────────────┘
        └───────┬───────┘                            │
                │ N                                  │
                ▼                                    │
        ┌───────────────┐                            │
        │      IS       │         N                  │
        │ THE MATERIAL  │────────────────────────────┤
        │   READILY     │                            │
        │  ACCESSIBLE?  │                            │
        │   See Note 3  │                            │
        └───────┬───────┘                            │
                │ Y                                  │
                ▼                                    │
        ┌───────────────┐                            │
        │      IS       │   N                        │
        │  THE DAMAGE   │─────────┐                  │
        │  EXTENSIVE?   │         │                  │
        └───────┬───────┘         │                  │
                │ Y               │                  │
                ▼          See Note 4                │
        ┌───────────────┐         │                  │
        │      IS       │   Y     ▼    ┌──────────────────┐
        │  SEALING OR   │───────────▶ │ SEAL OR ENCLOSE  │──┤
        │  ENCLOSURE    │              │   See Note 4     │  │
        │   FEASIBLE?   │              └──────────────────┘  │
        │   See Note 5  │                                    │
        └───────┬───────┘                                    │
                │ N                                          │
                ▼                                            ▼
        ┌───────────────┐                          ┌──────────────┐
        │    REMOVE     │                          │  MANAGEMENT  │
        │ See Notes 1 and 6│                       │  See Note 7  │
        └───────────────┘                          └──────────────┘
```

*Control of Asbestos at Work Regulations 1987

Chart 4: Abestos-cement products

Refer to Sections 4 and 5.

Notes

1. The chart deals with material which is considered not to be in good condition. If crocidolite asbestos is present, the HSE must be notified 28 days before work begins. Any work on asbestos-cement products should follow the guidance set out in HSE Guidance Note EH 36[16].

2. To be readily repairable damage to the material must be slight. Surface scratches may be sealed or painted, breaks may be taped and small punctures patched with filler.

3. The term accessible is explained in paragraph 4.11. Asbestos-cement is a very common material. It is unlikely to be sealed where it is used outside and where it is used inside buildings, sealing is likely to be confined to painting – although some products have factory applied coatings. Water damage and vermin are unlikely to be a problem, although the material becomes porous with age and may then allow water to leak through.

4. Accessible asbestos-cement which is not readily repairable but which has only suffered slight damage may be sealed with a suitable coating. The surface should be prepared and damaged material repaired if possible (see Note 2). Asbestos–cement products should not be sanded or wire-brushed. Dusty surfaces can be cleaned with a suitable industrial vacuum cleaner fitted with a high efficiency filter,[15] or wiped with a damp cloth (which should be sealed in a plastic bag whilst still damp). They must not be cleaned with a domestic vacuum cleaner. Asbestos-cement used outside may need treatment with a biocide before painting[17]. Asbestos-cement is alkaline and should be primed with an alkali-resistant primer, or a chlorinated rubber or oleo-resinous paint, followed by one or more top coats. Where possible both sides should be painted. Installations which are badly deteriorated and will not allow a surface coat to adhere should be removed.

5. It should not be necessary to attach warning labels to every asbestos-cement product that is found, particularly board that has been decorated and is in domestic premises. It is important that where replacement work is liable to cause frequent contact with the material, and where replacement or demolition is necessary, a note is made on plans and those who may have to disturb the material are made aware of its existence. Warning notes should be attached where material is readily accessible. In the case of asbestos-cement roofs, the notes should indicate the material is fragile, and the risk of falling through it. Asbestos-cement roofing sheets or tiles may have fibre washed off which can collect in gutters, and this should be borne in mind during maintenance of buildings. If the material is not readily accessible, then the management procedures given in paragraph 5.4 should be followed. In housing, occupants, especially those in public sector rented accommodation, should be made aware of the location of any asbestos materials and advised of appropriate precautions.

6. Removal (paragraph 5.7) of large amounts of asbestos-cement should be carried out by a specialist contractor or trained staff. Small quantities of asbestos-cement can be safely removed by householders, provided that safety precautions described in paragraph 3.12 and in the leaflet 'Asbestos in Housing'[39] are followed. It may not be necessary to empty buildings or seal off the working area during removal, but people not engaged in the work should be kept out of the working area. Sheets should be wetted to suppress dust and removed whole, not broken up. The material removed and any dust and debris should be carefully collected, smaller pieces dust dampened and sealed in strong plastic bags marked **'ASBESTOS'**. The whole area should be thoroughly cleaned (see Note 4) using a dustless method. After large scale work, especially where there has been breakage of asbestos-cement sheets, the airborne fibre concentration should be measured before the area is reoccupied.

Chart 4: Asbestos-cement products

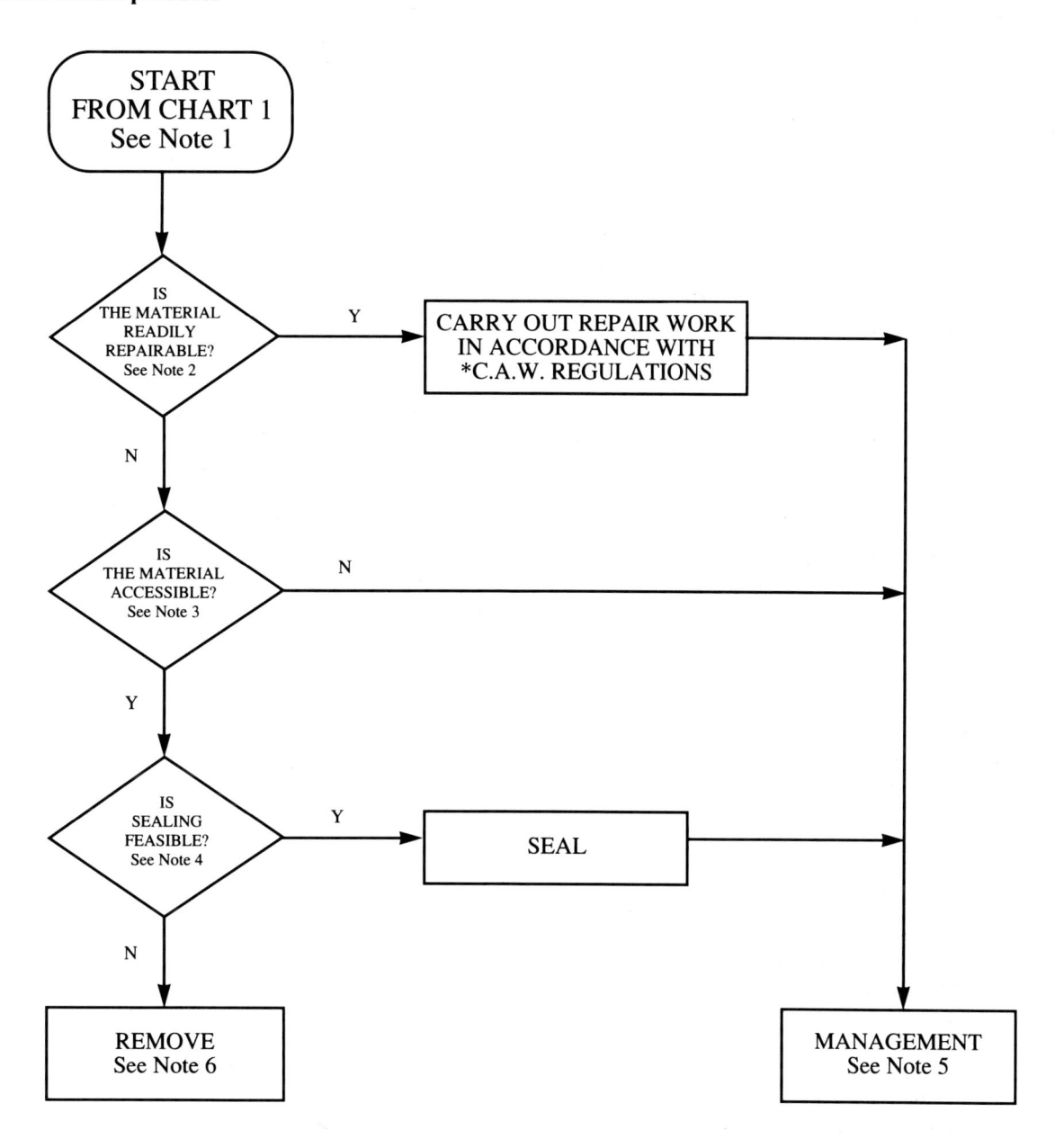

```
           ┌─────────────┐
           │   START     │
           │ FROM CHART 1│
           │  See Note 1 │
           └──────┬──────┘
                  │
                  ▼
            ╱ IS          ╲
           ╱ THE MATERIAL  ╲    Y      ┌──────────────────────┐
          ╱  READILY        ╲─────────▶│ CARRY OUT REPAIR WORK│
           ╲  REPAIRABLE?   ╱           │  IN ACCORDANCE WITH  │
            ╲ See Note 2   ╱            │  *C.A.W. REGULATIONS │
             ╲            ╱             └──────────┬───────────┘
                  │ N                              │
                  ▼                                │
            ╱ IS          ╲                        │
           ╱ THE MATERIAL  ╲    N                  │
          ╱  ACCESSIBLE?    ╲───────────────────────────────────▶
           ╲  See Note 3   ╱                       │
            ╲             ╱                        │
                  │ Y                              │
                  ▼                                │
            ╱ IS          ╲                        │
           ╱  SEALING      ╲    Y    ┌──────┐     │
          ╱  FEASIBLE?      ╲───────▶│ SEAL ├─────┤
           ╲ See Note 4    ╱         └──────┘     │
            ╲             ╱                        │
                  │ N                              ▼
                  ▼                         ┌──────────────┐
          ┌───────────────┐                 │  MANAGEMENT  │
          │    REMOVE     │                 │  See Note 5  │
          │  See Note 6   │                 └──────────────┘
          └───────────────┘
```

*Control of Asbestos at Work Regulations 1987

Annex 4

Notes on procedures for other asbestos materials

These notes refer to materials listed in paragraph 4.7 other than sprayed coating and lagging, insulating boards etc. and asbestos-cement products.

Millboard, paper and paper products, ropes, yarns and cloths. These materials have high asbestos content and except where bonded, for example as in gaskets, are friable and will release fibres easily. If damaged, deteriorating or releasing dust, or if they are installed in places where they are subject to abrasion and wear, they should be replaced with an asbestos-free substitute. Removal of unbonded, friable material should be undertaken by a specialist contractor, following procedures set out in the HSC Approved Code of Practice for Work with asbestos insulation and asbestos coating.[2] It may be necessary to empty the building or seal-off the working area if large quantities of the materials are involved. In any case, people not engaged in the work should be kept out of the working area. After removal, the whole area should be thoroughly cleaned up using a suitable industrial vacuum cleaner fitted with a high efficiency filter, or wiped with a damp cloth, which should be sealed in a plastic bag whilst still damp. Following extensive removal, the airborn fibre concentration should be measured using procedures specified in HSE Guidance Note EH 10[18] prior to reoccupation. This will probably not be necessary where only small quantities of material have been removed. Materials which are undamaged and are not likely to become damaged should be left alone and their condition checked periodically.

Bitumen roofing felts, damp-proof courses, semi-rigid asbestos-bitumen products and asbestos-bitumen coated metal. The asbestos fibres in these materials are firmly bound and are not likely to be released during installation and normal use, but bitumen materials may become brittle with age and break up. If this occurs, they should be removed carefully and any adhering material removed manually not by power grinding. Waste material should not be burned, neither should the asbestos-bitumen coating be burned off scrap metal sheet, as this will release asbestos fibres.

Asbestos-paper backed vinyl flooring, unbacked (homogeneous) vinyl flooring and floor tiles. Fibres bonded into homogeneous flooring will be released as it wears, but the rate of release will be very low except where the flooring is subject to heavy wear in commercial or industrial buildings. The condition of the flooring should be assessed. If it is badly worn it should be covered, or replaced, with an asbestos-free product. When flooring containing asbestos is being removed it should be lifted carefully and any residue left behind removed without abrasion. It is preferable to cover or skim the uneven surface. When removing asbestos-paper backed vinyl flooring (cushion flooring), any loose dust and any backing sticking to the floor should be dampened and carefully removed while damp. Backing must not be removed by power sanding. Waste materials must not be burned as this may release asbestos fibres.

Textured coatings, eg. 'Artex' and paints containing asbestos. These materials are unlikely to release asbestos fibres during normal use and it is preferable to leave them in place. When redecoration is undertaken, it is best to paint them over. If for any reason it is necessary to remove textured coatings and paints which contain asbestos, they must not be removed dry or sanded down as this will release asbestos fibres. Some paints and coatings can be removed by wetting them thoroughly and scraping them off as a slurry. Other paints and coatings may require treatment with paint remover prior to wetting and scraping off. If this is the case then precautions should be taken to prevent paint remover and material soaked in paint remover from entering the eyes, especially when removing materials from ceilings. If known, the manufacturer or distributor of the textured coating may be able to advise on procedures for its removal.

Sealants and mastics. These are unlikely to release asbestos fibres during normal use. They should not be sanded down with power tools as this will release asbestos fibres.

Asbestos reinforced plastics. These are unlikely to release asbestos fibres during normal use. If for any reason it is necessary to remove them, they should not be cut up with high speed power tools as this may release asbestos fibres.

Annex 5

The Asbestos Label

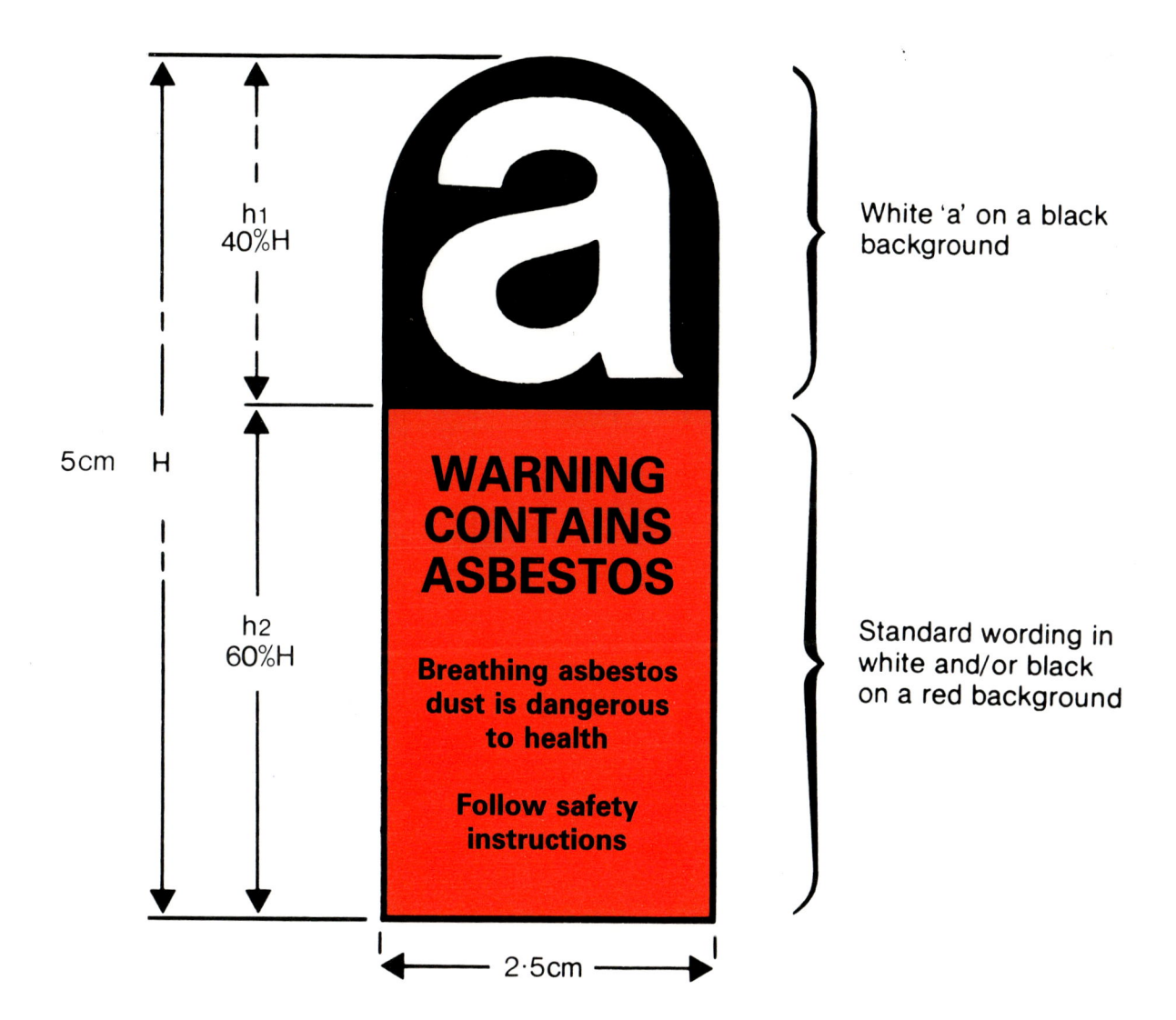

- 5cm H
- h1 40%H — White 'a' on a black background
- h2 60%H — Standard wording in white and/or black on a red background
- 2·5cm

WARNING CONTAINS ASBESTOS

Breathing asbestos dust is dangerous to health

Follow safety instructions

49

ANNEX 6: MODEL ASBESTOS SURVEY REPORT FORM

ADDRESS

SURVEY REFERENCE NO
SHEET NO.   TOTAL NO. SHEETS

DESCRIPTION OF PREMISES:-

TYPE   CONSTRUCTION
NO. OF ROOMS   NO. OF BEDROOMS
VACANT/OCCUPIED   DOMESTIC/INDUSTRIAL/OTHER
NUMBER OF OCCUPANTS   CHILDREN PRESENT? (Y/N)

SUSPECT MATERIAL IDENTIFICATION	1	2	3	4	5	6	7	8	9	10
LOCATION (ROOM) POSITION IN ROOM INTERNAL/EXTERNAL (INT/EXT) TYPE OF MATERIAL(1) FUNCTION OF MATERIAL SAMPLE TAKEN (Y/N) RESULT OF ANALYSIS(2) PHOTOGRAPH TAKEN (Y/N) **SUSPECT MATERIAL ASSESSMENT** CONDITION(3) ACCESSIBLE (Y/N) DUST RELEASE (Y/N) **SUGGESTED REMEDIAL MEASURES** NONE (MANAGEMENT) REPAIR SEAL OR ENCAPSULATE ENCLOSE REMOVE										

URGENT ACTION REQUIRED (Y/N)

GENERAL REMARKS

SURVEY OFFICER(S)

                                              DATE

(1) Spray coating or lagging (L); Insulating board or tile (I); Asbestos-cement sheet (AC); Other (O).
(2) Chrysotile (Ch); Amosite (A); Crocidolite (Cr); Other (O).
(3) Good (G); Slightly damaged (S); Badly damaged (B).

Annex 7

The Asbestos Waste Label

ASBESTOS WASTE CONTAINS	
ASBESTOS BLUE ASBESTOS BROWN 2212	ASBESTOS WHITE 2590

TOXIC BY INHALATION
POSSIBLE RISK OF IRREVERSIBLE EFFECTS
DANGER OF SERIOUS DAMAGE TO HEALTH BY PROLONGED EXPOSURE
WEAR PROTECTIVE CLOTHING
DO NOT BREATHE DUST
IF YOU FEEL UNWELL, SEEK MEDICAL ADVICE (SHOW LABEL WHERE POSSIBLE)

NAME, ADDRESS AND TELEPHONE NUMBER OF SUPPLIER:—

! DANGEROUS SUBSTANCE

Although two types of label are allowed for receptacles containing less than 25 kg of asbestos waste, the type of label illustrated, specified in Regulation 10 (2) (a) (i) of the Classification, Packaging and Labelling of Dangerous Substances Regulations 1984 (the CPL Regulations) is recommended, as this also applies to receptacles containing 25 kg or more. The label comprises the following.

i. The name and address of the manufacturer, importer, wholesaler or supplier of the substance.

ii. The designation or accepted common name of the substance, namely blue, white or brown asbestos. Asbestos waste is regarded as a preparation and it should be identified as: 'Asbestos Waste Contains Blue, Brown or White Asbestos', as the case may be.

iii. The risk phrases namely:
- R.23 toxic by inhalation;
- R.40 possible risk of irreversible effects;
- R.48 danger of serious damage to health by prolonged exposure.

The safety phrases, namely:
- S.22 do not breath dust;
- S.36 wear suitable protective clothing;
- S.44 if you feel unwell seek medical advice (show the label where possible).

Plus, as appropriate to the substance and its intended use:
- *either* S.35. This material and its container must be disposed of in a safey way,
- *or* S.38. In case of insufficient ventilation wear suitable respiratory equipment.

iv. The substance identification number given in Part 1A2 Approved List[42], ie. 2212 for blue and brown asbestos, and 2590 for white asbestos.

v. The hazard warning sign, ie: 'Dangerous Substance' (black on white diamond illustrated in Part 1 of Schedule 2 and specified in Part II of Schedule 2 of the CPL Regulations).

For receptacles containing more than 25 kg of asbestos, or substances and wastes containing it, items i. to v. above should be on the label, together with information on the nature of the dangers to which the substance may give rise and the emergency action to be taken, for example, in the event of spillage. However, this information may be shown on a separate statement accompanying the package, if that statement also shows the name and address of the consignor, the designation of the substance and the classification 'Dangerous Substance'.

The exact dimensions of labels and the method of marking or labelling packages are given in Regulation 13 of the CPL Regulations. The label can be indelibly marked on the package, or it will be sufficient if it is clearly and indelibly printed and securely fixed to the package, 'with its entire surface in contact with it'. The label must be easily readable, 'when it is placed in an attitude in which it may normally be expected to be placed'. The label can be on an inner receptacle, provided it can be clearly seen through any outer package. This allows clear outer sacks to be used in suitable circumstances. Once the EC has agreed a new supply label for asbestos, the specimen label and risk phrases at Annex 7 may have to be amended. Any labelling for the first time must refer to Part 1A of the current edition of the CPL Approved List.

References

[1] Health and Safety Commission. Asbestos: final report of the advisory committee. Volume 1 ISBN 0 11 883293 X. Volume 2 ISBN 0 11 883294 8*. HMSO 1979.

[2] Health and Safety Commission. Work with asbestos insulation, asbestos coating and asbestos insulating board. Approved Code of Practice (COP3) (Revised March 1988) ISBN 0 11 883979 9. HMSO March 1988.

[3] Council Directive 83/477/EEC on the protection of workers from the risks related to exposure to asbestos at work.

[4] Council Directive 83/478/EEC amending for the fifth time Directive 76/769/EEC on the approximation of the laws, regulations and administrative provisions of the Member States relating to restrictions on the marketing and use of certain dangerous substances and preparations.

[5] Health and Safety Executive. A guide to the Asbestos (Licensing) Regulations 1983. Health and Safety series booklet HS(R)19. (Revised 1989). ISBN 0 11 885489 5. HMSO 1989.

[6] Health and Safety Commission. Control of Asbestos at Work Regulations 1987. ISBN 0 11 078115 5 HMSO 1988 and the Approved Code of Practice (COP21) ISBN 0 11 883984 5.

[7] Acheson, E D and Gardner, M J. The control limit for asbestos. Health and Safety Commission. ISBN 0 11 883700 1. HMSO 1983.

[8] Doll, R and Peto, J. Effects on health of exposure to asbestos. Health and Safety Commission. ISBN 0 11 883803 2. HMSO 1985.

[9] National Research Council. Asbestiform Fibres – Non-occupational Health Risks. National Academy Press. ISBN 0 309 03446 9. Washington DC 1984.

[10] US Consumer Product Safety Commission. Chronic hazard advisory panel on asbestos. US Consumer Product Safety Commission. Washington DC 1983.

[11] Royal Commission on Matters of Health and Safety Arising from the Use of Asbestos in Ontario. Report. ISBN for complete set of three volumes 0 7743 8508 1. Ontario Ministry of the Attorney General, Toronto, Ontario 1984.

[12] Le Guen, J M and Burdett, G. Asbestos concentrations in public buildings – a preliminary report. Annals of Occupational Hygiene: 24, 2 1981.

[13] Sir Richard Doll and Professor Julian Peto. Private Communication.

[14] Conway, D M and Lacey, R F. Asbestos in drinking water: Results of a survey. Technical Report TR 202 Water Research Centre 1984.

[15] Health and Safety Executive. Work with asbestos insulating board. EH 37 (Revised December 1989) ISBN 0 11 885423 2. HMSO 1989.

[16] Health and Safety Executive. Work with asbestos cement. EH 36 (Revised December 1989). ISBN 0 11 885422 4. HMSO 1989.

[17] Building Research Establishment. Control of lichens, moulds and similar growths. Digest No 139. ISBN 0 11 726857 7. Dept. of the Environment, 1982.

[18] Health and Safety Executive. Asbestos: exposure limits and measurement of airborne dust concentrations. Guidance Note EH 10. ISBN 0 11 885552 2. (Revised March 1990).

[19] Health and Safety Executive. Asbestos fibre in air – Light microscopic methods for use with the control of Asbestos at Work Regulations. MDHS 39/3 ISBN 0 71776 03563. HMSO June 1990.

[20] Fisher, R W and Rogowski, B F W. Results of surface spread of flame tests on building products. Building Research Establishment Report. ISBN 0 11 670542 6*. HMSO 1976.

[21] Department of Education and Science. Maintenance and Renewal in Educational Buildings – Needs and Priorities. Department of Education and Science Architects and Building Group Design Note 40. ISSN 0141 2825. 1985.

[22] Health and Safety Executive. Health and Safety in demolition work, Guidance Note GS 29/1, Part 1: preparation and planning; ISBN 0 11 885405 4. 1988 Guidance Note GS 29/2, Part 2: legislation; ISBN 0 11 883589 0*. 1984 Guidance Note GS 29/3. Part 3: techniques; ISBN 0 11 883609 9. 1984 Guidance Note GS 29/4. Part 4: health hazards; ISBN 0 11 883604 8. 1985.

[23] Department of the Environment. Waste Management Paper No 23. Special wastes: a technical memorandum providing guidance on their definition. ISBN 0 11 751555 8*. HMSO 1981.

[24] Department of the Environment: P Circular 4/81. Welsh Office: Circular 8/81. Control of Pollution Act 1974 Control of Pollution (Special Waste) Regulations 1980. ISBN 0 11 751510 8.*

[25] Hodgson, A A. Alternatives to Asbestos – the Pros and Cons. Editor A A Hodgson. Clinical Reports on Applied Chemistry Volume 26. 1989. Society of Chemical Industry, John Willey and Sons. ISBN 0 471 92353 2.

[26] Department of the Environment. Waste Management Paper No 18. Asbestos wastes. ISBN 0 11 751384 9. HMSO 1979.

[27] Department of Trade Circular No 1 Imported Catalytic Gas Heaters. 22 February 1983.

[28] The Gas Catalytic Heaters (Safety) Order 1983 (Statutory Instrument 1983 No 1696).*

[28A] The Gas Catalytic Heaters (Safety) Regulations 1984 (Statutory Instrument 1984 No 1802).

[29] Prentice, J and Keech, M. Alteration of Asbestos with Heat, Microscopy and Analysis: March 1989.

[30] The Electricity Council (now the Electricity Association). Guidance to local authorities on asbestos in some electric storage heaters. The Electricity Council. December 1984.

[31] Pye, A M. Alternatives to asbestos in industrial applications. Asbestos Volume 1: Properties, Applications and Hazards. Edited by L Michaels and S S Chissick. Wiley Interscience pp 339 to 373 ISBN (Vol. 1) 0 471 99798 X. 1979.

[32] Health and Safety Executive. Alternatives to asbestos products: A review. ISBN 0 11 883812. HMSO 1986.

[33] Hodgson, A A. Alternatives to asbestos products. Anjalena Publications Ltd. (Second edition). ISBN 0 95 10148 2. 1987.

[34] Stanton, M F and Wrench C. Mechanisms of mesothelioma induction with asbestos and fibrous glass. Journal of the National Cancer Institute. 1972, 48, 797–821.

[35] World Health Organizations. Biological effects of man-made mineral fibres. Report on a WHO/IARC meeting. Copenhagen April 1982. Euro Reports and Studies 81. World Health Organization. ISBN 9 28 901247 1. Regional Office for Europe 1983.

[36] World Health Organization. Biological effects of man-made mineral fibres. Proceedings of a WHO/IARC Conference, in association with JEMRB and TIMA, Copenhagen, April 1982. Volume 1 Introduction and Sessions I–V, ISBN 92 890 1026 6. Volume 2 Sessions VI–IX and annexes, ISBN 92 890 1025 8. Nonserial publications of the WHO Regional Office for Europe 1984.

[37] Health and Safety Commission. Man-made mineral fibres. Report of a working party to the Advisory Committee on Toxic Substances. ISBN 0 11 883251 4*. HMSO 1979.

[38] Health and Safety Executive. Man-made mineral fibres Guidance Notice EH 46. ISBN 0 11 885571 9. HMSO November 1990.

[39] Department of the Environment. Asbestos in Housing. December 1986.

[40] Burdett, G J and Jaffrey, S A M T. 'Airborne asbestos concentrations in buildings'. Ann. Occup. Hyg. 30, 185–199 (1986).

[41] Rood, A P 'Environmental Asbestos'. Research Paper 25 (Health and Safety Executive, 1989) ISBN 0 71 760313 X.

[42] Health and Safety Executive. Information Approved for the Classification, Packaging and Labelling of Dangerous Substances for Supply and Conveyance by Road (Authorised and Approved List) 3rd Edition. ISBN 0 11 885542 5. HMSO.

[43] Health and Safety Executive. Enclosures provided for work with asbestos insulation, coatings and insulating board. Guidance Note EH 51. ISBN 0 11 885408 9. HMSO. January 1989.

[44] ICRCL 64/85. Asbestos on Contaminated Sites. 2nd edition October 1990. DOE Publication Sales Unit, Building 1, Victoria Road, S Ruislip, Middlesex HA4 ON2.

[45] Council Directive 85/610 EEC amending for the Seventh Time Directive 76/769/EEC on the approximation of the laws, regulations and administrative provisions of the Member States relating to restrictions on the marketing and use of certain dangerous substances and preparations.

[46] Council Directive 87/217 EEC on the prevention and reduction of environmental pollution by asbestos.

[47] Health and Safety Executive. Probable asbestos dust concentrations at construction processes. EH 35 (revised) 1989. ISBN 0 11 885421 6. HMSO.

[48] Institute of Wastes Management: Code of Practice for the Disposal of Asbestos Waste. October 1988. ISBN 0 90 294417 7.

---

* These publications are out of print, but copies can be obtained from reference libraries